Pandian Vasant, J. Joshua Thomas, Vladimir Panchenko, and Gerhard Wilhelm Weber

Smart Green Energy Production

De Gruyter Series on Smart Computing Applications

Edited by
Pandian Vasant, J. Joshua Thomas, Vladimir Panchenko, and
Gerhard Wilhelm Weber

Volume 11

Smart Green Energy Production

Edited by
Pandian Vasant, J. Joshua Thomas, Vladimir Panchenko, and
Gerhard Wilhelm Weber

DE GRUYTER

Editors
Dr. Pandian Vasant
TON DUC THANG UNIVERSITY
19 Nguyen Huu Tho Street
Tan Phong Ward, District 7
Ho Chi Minh City, Vietnam
pvasant@gmail.com

Prof. Dr. J. Joshua Thomas
UOW Malaysia KDU Penang University College
32, Anson Road
George Town
10400 Penang
Malaysia
joshua.j.thomas@gmail.com

Dr. Vladimir Panchenko
Russian University of Transport
Minayevskiy Pereulok 2
127994, Moscow,
Obraztsova street, 9
Russia
pancheska@mail.ru

Prof. Dr. Gerhard Wilhelm Weber
Faculty of Engineering Management
Poznan University of Technology
Plac Marii Skłodowskiej-Curie 5
60-965 Poznan
Poland
gerhard-wilhelm.weber@put.poznan.pl

ISBN 978-3-11-071448-7
e-ISBN (PDF) 978-3-11-071765-5
e-ISBN (EPUB) 978-3-11-071774-7
ISSN 2700-6239

Library of Congress Control Number: 2024941482

Bibliographic information published by the Deutsche Nationalbibliothek
The Deutsche Nationalbibliothek lists this publication in the Deutsche Nationalbibliografie; detailed bibliographic data are available on the Internet at http://dnb.dnb.de.

Cover image: monsitj/iStock/Getty Images Plus
Typesetting: Integra Software Services Pvt. Ltd.
Printing and binding: CPI books GmbH, Leck

www.degruyter.com
Questions about General Product Safety Regulation:
productsafety@degruyterbrill.com

MIX
Papier | Fördert gute Waldnutzung
FSC® C083411

Contents

List of authors

A. Jumayev
N. Allanazarov
B. Reyimov
K. Saryyev
M. Orazberdiyeva
A. Matyakubov
A. Matyakubov
S. Nazarov
Fazıl Gökgöz and Engin Yalçın
Er. Aditya Singh
JORGE M. DE SOUZA
FABRÍCIO CRISTÓFANI
GIOVANNI M. DE HOLANDA
RAFAEL TORRES-ESCOBAR
IGOR LITVINCHEV
PANDIAN VASANT
Panyagina A.E.
Cinzia Bellone
Pietro Ranucci
Valina Geropanta
C. Bellone
P. Ranucci
Riyadh Bouddou
Electrical Engineering, IRECOM Laboratory,
University of Djillali Liabes
Jabulani Monchwe
Elias Munapo
Martin Chanza

Chapter 1
Aganiyaz Jumayev
Bayramhan Street 62
Mary
Turkmenistan
a.jumayev_tm@mail.ru

Chapter 1
Allanazarov Nurmuhammet
Bayramhan Street 62
Mary
Turkmenistan
nurysh90@bk.ru

Reyimov Begench
Bayramhan Street 62
Mary
Turkmenistan

Chapter 2
Saryyev Kakageldi
Bayramhan Street 62
Mary
Turkmenistan
kakageldisaryyev@gmail.com

Chapter 3
Orazberdiyeva Mahriban
Bayramhan Street 62
Mary
Turkmenistan
mahribanorazberdiyeva@gmail.com

Chapter 4
Matyakubov Amirhan
Bayramhan Street 62
Mary
Turkmenistan
Amirhan31071989@mail.ru

Chapter 5
Nazarov Serdar
Bayramhan Street 62
Mary
Turkmenistan
marytdei@mail.ru

Chapter 6
Saliha Demirbüken
Department of Mathematics
Gebze Technical University
41400 Kocaeli
Turkey
sdemirbuken@gtu.edu.tr

Vilda Purutçuoğlu
Department of Statistics
Middle East Technical University
06800 Ankara
Turkey
vpurutcu@metu.edu.tr

https://doi.org/10.1515/9783110717655-203

Chapter 7
Er. Aditya Singh
School of Civil Engineering
Lovely Professional University
Phagwara, Punjab
India
aditya7777777k@gmail.com

Chapter 8
Vilda Purutçuoğlu
Department of Statistics
Middle East Technical University
06800 Ankara
Turkey
vpurutcu@metu.edu.tr

Başak Bahçivancı
Earlham Institute
Norwich Research Park
Norwich NR47UZ
UK
basakbahcivanci@gmail.com

Chapter 9
Mehmet Ali Kaygusuz
Department of Statistics
Middle East Technical University
06800 Ankara
Turkey

Vilda Purutçuoğlu
Department of Statistics
Middle East Technical University
06800 Ankara
Turkey

Chapter 10
Cinzia Bellone
Guglielmo Marconi University
Rome
Italy

Pietro Ranucci
Roma Tre University
Rome
Italy

Valina Geropanta
Greece Founder of Avrions
Technical University of Crete
Crete
Greece
c.bellone@unimarconi.it

Chapter 11
Ekin Can Erkuş
Intelligent Application DC
Huawei Technologies Co. Ltd. Turkey R&D Center
Istanbul
Turkiye

Vilda Purutçuoğlu
Department of Statistics
Middle East Technical University
06800 Ankara
Turkey

Chapter 12
Jabulani Monchwe
Department of Statistics and Operations Research
School of Economic Sciences
North-West University
Mafikeng
South Africa

Elias Munapo
Department of Statistics and Operations Research
School of Economic Sciences
North-West University
Mafikeng
South Africa
Elias.Munapo@nwu.ac.za

Martin Chanza
Department of Statistics and Operations Research
School of Economic Sciences
North-West University
Mafikeng
South Africa

Chapter 13
Khadija Rahal
High School of Sciences and Technology of Hammam Sousse (ESSTHS)
University of Sousse (US)
4011 Street Lamine Abassi Hammam Sousse
Tunisia
and
Laboratory of Electro-Mechanic Systems (LASEM)
National Engineering School of Sfax (ENIS)
University of Sfax (US)
B.P. 1173
km 3.5 Soukra
3038 Sfax
Tunisia
khadija.rahal@enis.tn

Zied Driss
Laboratory of Electro-Mechanic Systems (LASEM)
National Engineering School of Sfax (ENIS)
University of Sfax (US)
B.P. 1173
km 3.5 Soukra
3038 Sfax
Tunisia

Mohamed Salah Abid
Laboratory of Electro-Mechanic Systems (LASEM)
National Engineering School of Sfax (ENIS)
University of Sfax (US)
B.P. 1173
km 3.5 Soukra
3038 Sfax
Tunisia

N. Allanazarov and B. Reyimov

Chapter 1
Use of digital systems in the design system of photovoltaic solar stations

Abstract: Benefiting from digital systems in the introduction of "green" technologies in our country, creation of digital databases and designing of programs for undertaking projects will greatly assist in providing solutions for technological problems. Furthermore, there will be vast opportunities to develop these programs and databases.

With the purpose of providing the quality and reliability of tasks, an application program "Digital system of designing photoelectric solar station" is developed for designing the photoelectric solar station and its elements. This application program is completely developed in the language of "Delphi programming."

In the developed application program "Digital system of designing the wind power stations," it is considered the features of undertaking the project works in connection with designing the wind power station in a convenient and reliable way, making calculations for discovering the wind energy resources in the region and storing it in its database.

Keywords: Solar energy, solar radiation intensity in Turkmenistan, digital system of designing photoelectric solar station, wind energy, project wind energy station, digital system of designing the wind power stations

1.1 Introduction

Development of digital education is a modern administration system. It gives an opportunity for high-quality works in scientific production centers. Achievements in creating highly effective project systems are directly related to the methodological supply, analyses conducted in the initial stage of operations, precise conduction of projects and deep study of reasonable connection of its statistics.

An application program "Digital system of designing photoelectric solar station" allows to design photoelectric stations, a precise and brief performance of calculations in project works. It is also considered the features of storing performed calculations in its own base.

N. Allanazarov, Bayramhan Street 62, Mary, Turkmenistan, e mail: nurysh90@bk.ru
B. Reyimov, Bayramhan Street 62, Mary, Turkmenistan

https://doi.org/10.1515/9783110717655-001

In the developed application program "Digital system of designing the wind power stations," it is considered as the features undertaking the project work in connection with designing the wind power station in a convenient and reliable way, making calculations for discovering the wind energy resources in region and storing it on its database.

1.2 Literature review

Awasthi et al. [1] presented an introduction to solar photovoltaic cells and the materials used in their construction. They also discussed the types of solar photovoltaic systems and solar tracking systems. It focuses on the design and performance analysis of various biaxial solar tracking systems proposed in recent years.

Mohanty et al. [2] presented the prediction of solar radiation data using different soft computing approaches like multilayer perceptron, artificial neurofuzzy inference system and radial basis function for the east coast region of India. Based on the input data such as sunshine duration, temperature and humidity for the period 1984–1999, soft computing models for solar radiation data are analyzed.

Hold et al. [3] proposed a learning design for a course in cyber-physical assembly systems design which will be offered within the environment of the Industry 4.0 Pilot Factory at TU Wien, where students can learn the basics of cyber-physical assembly systems design on both theoretical and practical levels.

The objective of constructing a robot station in [4] is to allow students to design and produce a feasible-to-build mechatronic device using robust, low-cost components. It is a tool for students to gain experience by integrating different mechatronic fields of knowledge, as well as practicing the procedures needed to successfully accomplish their own design. The project starts by describing the target requirements to be achieved by the prototype robot, and these requirements will serve as the guideline for the design and further manufacture and testing of the system.

Vasar et al. [5] presented an efficient control approach for generators operating in variable conditions, based on measurements determined in the unload regime. This requires the permanent control of wind turbine speed and torque combined with the generator load adaptation.

There is also software "RETScreen" [6]. RETScreen is a clean energy management software system for energy efficiency, renewable energy and cogeneration project feasibility analysis as well as ongoing energy performance analysis. On the internet RETScreen Expert, an advanced premium version of the software is available in Viewer mode completely free of charge.

1.3 Digital system of designing photoelectric solar station

It is being taken into consideration to develop the digital education system, to create and extend the technological renewal in this area, to accelerate the technological development of educational institutions, to establish an important electronic database for all levels of education and also to have access to it through internet.

Based on the above-mentioned, "Conception of developing Digital educational system in Turkmenistan" has been adopted by the president of Turkmenistan in accordance with the decree no. 340 on September 15, 2017.

On the basis of leading achievements of world's digital technology, our president has given significant instructions to the national education system in order to modernize the educational system of our country and to be able to manage the modern technologies of every citizen of our country, to carry out considerable reforms on improving the training of educational personnel in using innovative approaches of digital education, modern technical devices, systems and databases.

For the stable development of economy in our country, it is significant to provide highly effective use of opportunities of energy sector and natural energy sources. This increases the balance of renewable and extraordinary energy sources, types of fuel options and secondary energy means in energy balance of Turkmenistan. This also provides the safety of health and living conditions of population and also an environment during the use of energy means.

The weather and geographical conditions of Turkmenistan give an opportunity to consume the renewable energy sources of our country in a modern way. Obviously, it helps to improve the energy supply of rural population and regions remote from cultural zones. Solar energy is one of the renewable energy sources. The weather and geographical conditions of Turkmenistan give an opportunity to generate solar energy and to consume it actively on industrial zones for power and heat energy [7]. In addition, it is deemed to use wind, geothermal and biofuel energies. Several works are regularly being carried out to develop and place the innovative technology, to produce renewable energy resources and to use the modern energy saving devices, tools and materials in the economic and social sectors of Turkmenistan.

It is required to follow the installation operation procedures that are set for drafting the project in order to provide the reliable and effective operation of photoelectric solar stations. Reliable and effective operation of photoelectric solar stations is related to factors such as the geographic coordinates of installation location of station, the values of the intensity of solar radiation reaching the surface of solar panels inclined at a suitable angle in accordance with the horizontal plane throughout the year, the inclination angles of solar panels for the respective location and the weather temperature. In accordance with the aforesaid, it is one of the major issues to determine the annual production capacity of the photoelectric solar station, to specify the precise location where it is planned to be installed and also to study the influence of weather conditions on the energy parameters of the station [7, 8].

Nowadays, because of the increasing demand for renewable energy sources across the world, great attention is paid to generate power energy on account of photoelectric power batteries with the purpose of environmental protection and introduction of "green" technologies in our country. It has a diversified function of research institutions in our country as it is a complex issue to create a common system for realizing special projects on the basis of information technologies, and requires solving numerous organizational and technological issues.

To advance the exploitation of renewable energy sources and introduce innovative technologies across Turkmenistan's economic sectors, as well as to utilize the latest energy-saving devices and materials in social sectors, the "State Program for Energy Savings 2018–2024" is being implemented. This program was established by a presidential decree issued on February 21, 2018. "Drafting the proposals on selection of locations for installing solar, wind energy complexes and also solar and wind cadasters in the territory of Turkmenistan and assessment of solar and wind energy sources between the years of 2018–2021." In addition, the president of Turkmenistan on April 12, 2019, passed the framework of conception of improvement of areas of "Altyn asyr" Turkmen lake in 2019–2025 (first stage 2019–2022; second stage 2022–2025) in accordance with the decree no. 1207. In the first stage, in order to protect the environment and to introduce the ecological "green" technologies in our country, it is planned to construct "Solar and wind driven power stations with capacity of 10 MW" [9].

If the solar panels are placed at an angle β in accordance with the horizontal plane, then the final amount of solar energy reaches the inclined surface on one sunny day in an average month as follows:

$$E_n = R \cdot E \tag{1.1}$$

where E is the final amount of solar energy reaches the inclined surface during a day in an average month; R is the ratio of solar radiations reach the surface that is declined to the determined angle on horizon and horizontal surface during a day in an average month.

The calculation coefficient of passing to the declined surface directed to the south from the horizontal surface equals the sum of the three solar radiations: direct, dispersed and reflected:

$$R = \left(1 - \frac{E_p}{E}\right) \cdot R_n + \frac{E_p}{E} \cdot \frac{1 + \cos\beta}{2} + \rho \cdot \frac{1 - \cos\beta}{2} \tag{1.2}$$

where E_p is the amount of dispersed solar radiation that reaches the horizontal surface during a day in an average month; E_p/E is the part of dispersed solar radiation that reaches the horizontal surface during a day in an average month; R_n is the calculation coefficient of passing to the declined surface directed to the south from the horizontal surface; β is the inclination angle of solar panel roof according to the horizon;

ρ is the reflection coefficient (albedo) of the Earth's surface and the surrounding objects, which is calculated as 0.7 in winter and 0.2 in summer.

In the calculation of solar direct radiation for an average month, the calculation coefficient of passing the declined surface directed to the south from the horizontal surface is determined as follows:

$$R_n = \frac{\cos(\varphi - \beta)\cdot\cos\delta\cdot\sin\omega_{3H} + \frac{\pi}{180}\cdot\omega_{3H}\cdot\sin(\varphi-\beta)\cdot\sin\delta}{\cos\varphi\cdot\cos\delta\cdot\sin\omega_3 + \frac{\pi}{180}\cdot\omega_3\cdot\sin\varphi\cdot\sin\delta} \tag{1.3}$$

where φ is the geographical latitude of location; β is the inclination angle of solar panel roof according to the horizon, in degrees; δ is the solar inclination angle that is between the straight line that joins the center of the Earth and the Sun and projection of this straight line onto the equatorial plane, in degrees (vertical angle of the Sun according to the equatorial plane):

$$\delta = 23,45\cdot\sin\left(360\cdot\frac{284+n}{365}\right) \tag{1.4}$$

where n is the order number of days in a year, a list starts from January 1 (serial number of days that an average calculation is undertaken for each day of a year). You can get the values of β, the solar inclination angle, from Table 1.1.

Table 1.1: Solar inclination angle.

Months	I	II	III	IV	V	VI	VII	VIII	IX	X	XI	XII
n	17	47	75	105	135	162	198	228	258	288	318	344
δ, degree	−20,9	−13	−2,4	9,4	18,8	23,1	21,2	13,5	2,2	−9,6	−18,9	−23

ω_3 is the hour angle of sunset (sunrise) for the horizontal surface.

$$\omega_3 = \arccos(-tg\varphi\cdot tg\delta) \tag{1.5}$$

ω_{3H} is the hour angle of sunset for the declined surface directed to the south

$$K = \omega_{3H} = \arccos[-tg(\varphi-\beta)\cdot tg\delta)] \tag{1.6}$$

The following information is provided in Table 1.2: the geographical coordinates of cities of regions of Turkmenistan and values of intensity of solar radiation hit the solar panel surface declined to the β optimal angle in accordance with the horizontal plane located in these cities and also the optimal inclination angle for these cities. The intensity information of solar radiation that hits the surface of solar panels inclined to the β angle according to the horizontal plane of regions of Turkmenistan is shown in Figure 1.1. The intensity of solar radiation that hits the surface of solar panels inclined to the optimal angle according to the horizontal plane in the territory of Turkmenistan is changing from the value of 1,819,882 kW · h/m^2 ann (Balkanabat city) to the value of 1,897,407 kW · h/m^2 ann [9].

The proposed application program gives an opportunity to perform the following operations and services:

- Stores the database of technical specifications of exploited devices (photoelectric solar panels, observation devices for accumulator batteries, accumulator batteries, and invertors) while the designing operations of photoelectric solar station are being carried out. Additionally, there are features of inserting the updated technical specifications of these devices into the database and removing the devices that are not in use.
- There is a feature of storing conducted calculations in an application program in its own database.
- There is a feature of printing the conducted calculations with its explanation.
- It is also considered to include the features of loading the conducted calculations with their explanations into storage in the PDF and DOC file formats.

By entering the main window, the following operations can be done:

- To calculate the value of solar radiation that freely reaches the plane surface
- Calculation of photoelectric solar stations
- Database of accumulator batteries
- Database of photoelectric solar stations
- Database of controller
- Database of invertors

Calculations begin by inserting the initial information about the place where the project work is going to be performed.

In this chapter, there is a notion of calculating the value of free solar radiation that reaches the plane surface. It is necessary to define the α-angle of solar radiation that hits the surface inclined horizontal and also the definite angle for calculating the solar radiation that reaches the surface inclined to the definite angle according to the horizon at the determined location. At the moment of determination, it is related to the location of point A in accordance with Sun's rays and L-geographic latitude of its location, time angle τ and inclination angle. If those parameters are known, it has an opportunity to calculate the value of free solar radiation that reaches the plane surface (Figure 1.1) [10].

1.3.1 Notion about calculations of photoelectric solar station

In this chapter, select the table of calculation of photoelectric solar station from the menu of "Calculations" in the main window.

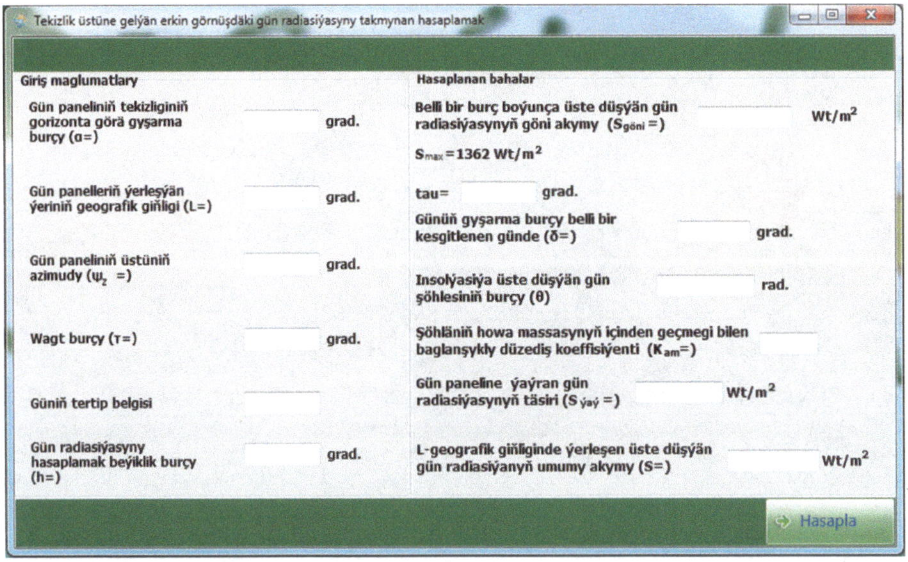

Figure 1.1: Calculation of the value of free solar radiation reaching the plane surface.

Calculation begins by selecting the table of "Start calculation" from the recently opened window "Operations" or by harmoniously pressing "Ctrl + N" on the keyboard [11].

In the new opened window, list of loads of consumers is inserted (Figure 1.2) and then the calculations are continued by inserting the initial information about the location where photoelectric solar station is planned to construct. While the calculations are performed, the features of selecting the placement of photoelectric solar station placement are also considered. It means that it gives an opportunity to take into account the total area of photoelectric solar station on the basis of calculation performed on consuming capacity (Figure 1.3).

As a result of calculations, it is considered to get detailed information about the capacities and numbers of accumulator batteries, solar batteries, invertors, and controllers that are necessary for photoelectric solar station to generate average power energy in a month. The performed calculations are stored in the database. In case of necessity, it is considered the features of reviewing the information and sending them to the storage hardware in PDF and Doc file formats. It has an opportunity of printing the calculations too.

1.3.2 Menu of database

A database consists of information complex with settings related to one definite way. It is an application program tool for creating the new database structure, controlling the database system, completing with the contents, editing the contents and creating

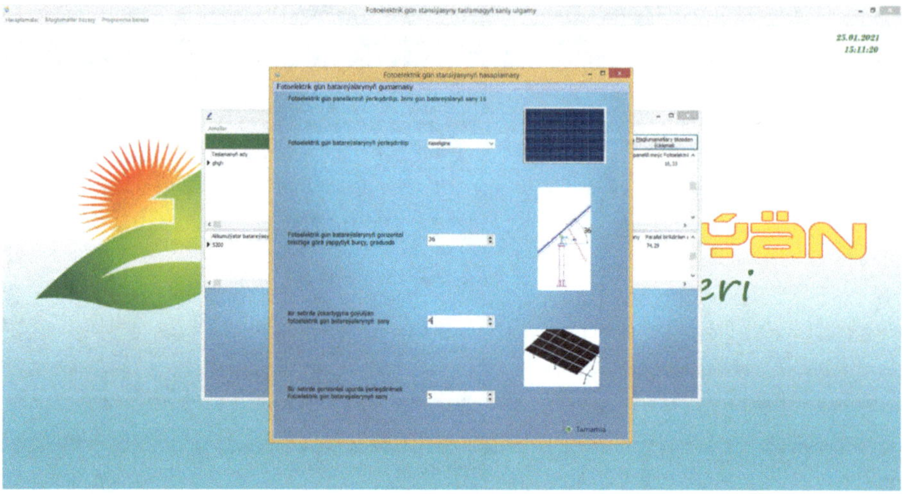

Figure 1.2: Window for inserting the loads of consumers.

Figure 1.3: Window for selecting the placement of solar panels.

the different consumer information (selecting, organizing, filtering, etc.). This database stores the information of technical specifications of devices (accumulator batteries, photoelectric solar batteries, invertors and controllers) which are going to be used during the calculations (Figure 1.4). It is available to load the new information into the database, edit the loaded information and remove the information out of use.

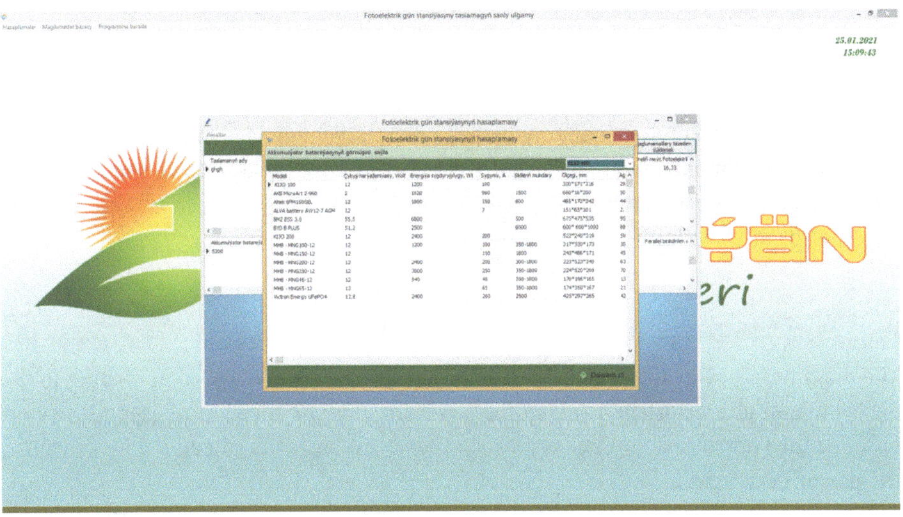

Figure 1.4: Window for selecting the accumulator batteries.

1.4 Digital system of designing the wind power station

One of the major types of alternative energy sources is wind energy. As per the assumptions of world scholars, the overall potential amount of wind energy sources distinguishes with its efficiency in all six continents. In connection with the installation condition of wind power station, the regular supply of stations with wind energy sources and measurement of the wind flow speed are defined. Thus, small increasing of an average speed of wind flow leads to considerable increase of production amount of power energy. Therefore, this indicator (an average speed of wind flow) has a great impact on the financial and economic efficiencies of the project.

For supplying power energy to consumers, the level of consuming the renewable energy sources, using the wind power stations in conversion of wind energy to power energy and putting it into operation is chosen in connection with complex factors. These significant symptoms include the energy potential of wind resources and its alteration according to the time, capacity and energy importance at the installation site, demands for reliability of the power supply system and the technical-economic statistics of the power supply system.

Digital economy is accepted as an overall system of economic and social activities.

Major advantages of digital economy are as follows:

- Increasing the labor productivity
- Decreasing the production expenses
- Increasing the ability to competition
- In a short time, precise conduction of calculations and projects is carried out by the digital system

The introduction of digital economy leads to digital technologies for the long-term development of economy and high productivity, to enhance the quality of social services and to establish the ideal types.

The speed variation was considered as

$$V(t) = 12 - 5 \cdot \frac{2\pi}{40} t \tag{1.7}$$

The maximum power that can be extracted for a certain wind speed according to the experimental characteristic is given for a certain angular speed ω_{ref}, used as a reference for the control system. The reference speed can be easily calculated by canceling the power derivative:

$$\frac{d(P_{\text{WT}})}{d\omega} = 0 \tag{1.8}$$

Considering the mechanical equation of the wind turbine

$$J\frac{d\omega}{dt} = M_{\text{WT}} \tag{1.9}$$

where J is the equivalent inertia and M_{WT} is the wind turbine torque.

For $dt = \Delta t = 500$ [s] and $d\omega = \Delta\omega = \omega_k - \omega_{k-1}$, the resulting turbine power is

$$P_{\text{WT}} = J\frac{d\omega}{dt} \cdot \omega = 50\frac{\omega_{k-}\omega_{k-1}}{500} \cdot \omega_k^* = P_{\text{ECH}} \tag{1.10}$$

where ω_k is the angular speed at moment t_k and ω_k^* is the angular speed at time $(t_{k-1} + t_k)/2$.

To ensure the quality and reliability of prior tasks, it stands as a major problem to provide fast and reliable designing of wind power stations in institutes and scientific research centers that undertake special project works. At this point of view, an application program, "Digital system of designing the wind power stations" is developed. An offered application program is completely developed in the programming language of C sharp (C#) [11].

Figure 1.5: Main window of the application program for designing the digital system of a wind power station.

This application program in Figure 1.5 has initiated by carrying out the following operations by entering the main window.

1.4.1 Database section

The database section contains information about technical specifications of wind power stations, annual average wind speeds in the selected areas and their blowing directions. It is also considered to insert the new technical specifications of wind power stations to the database by new achievements of world experience and to remove the information out of use from database (Figure 1.6). Moreover, the information about the winds according to the areas is inserted into the database. Further, it is possible to download information about winds as "*xls*" files from the satellite data system and to use the downloaded information widely for conducting the calculation and analysis (Figure 1.7) [10].

1.4.2 Calculation section

The calculation section includes several subsections:

1. Productivity of wind power station
2. Calculation on load

3. Hourly productivity of wind power station
4. Calculations performed
5. Calculation performed on loads

Figure 1.6: Window for downloading the technical specifications of wind power stations to the database.

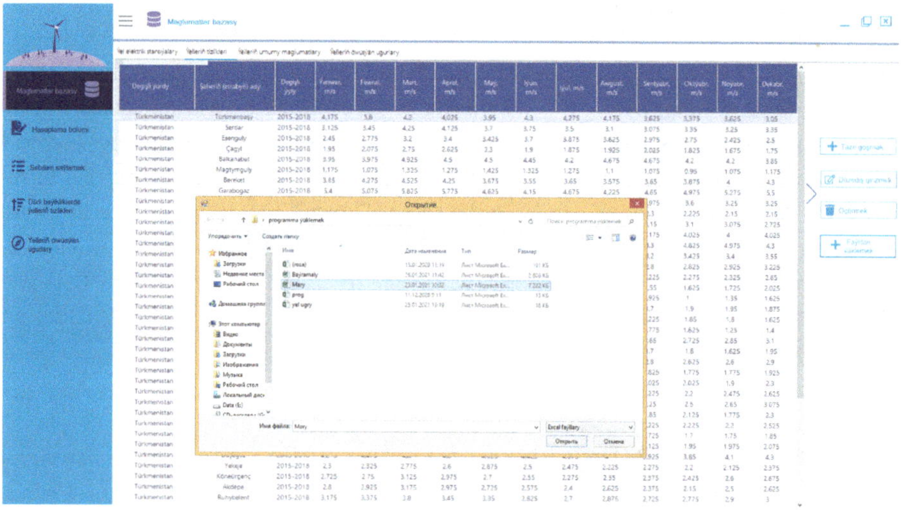

Figure 1.7: Window for working with information about winds on areas.

The productivity of wind power station allows to determine an hourly, a daily and a monthly production capacity of wind power station and to draw up the comparison graph of productivity of power energy (Figure 1.8). It is possible to store the calculations in the database.

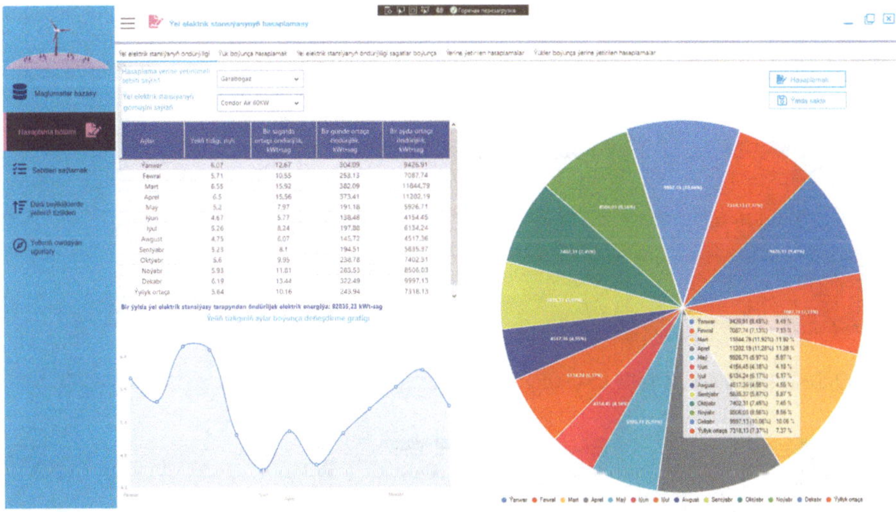

Figure 1.8: Window for determining the productivity of wind power station.

In the calculation window on loads, calculations are done by providing one of the any objects with power energy on account of independent wind power station. In this section, in case of inserting the capacity, exploitation date and number of devices consuming the power energy in the realization area of project, an application program defines the consumed power energy during day and night. In the project work, it determines the necessary number and productivity of wind power station for the region (Figure 1.9). In consequence, calculations are stored in the database.

1.4.3 Hourly productivity of wind power stations

In this section, hourly power energy productivity of wind power stations can be analyzed for regions stored in the database (Figure 1.10).

Figure 1.9: Calculation window on loads.

Figure 1.10: Hourly productivity window of wind power stations.

1.4.4 Calculations performed

In this section, it is taken into account to access calculations, which are determined in the wind power station productivity section and stored in the database and print them out or export them into the data storage devices in pdf, doc and xls file formats (Figure 1.11).

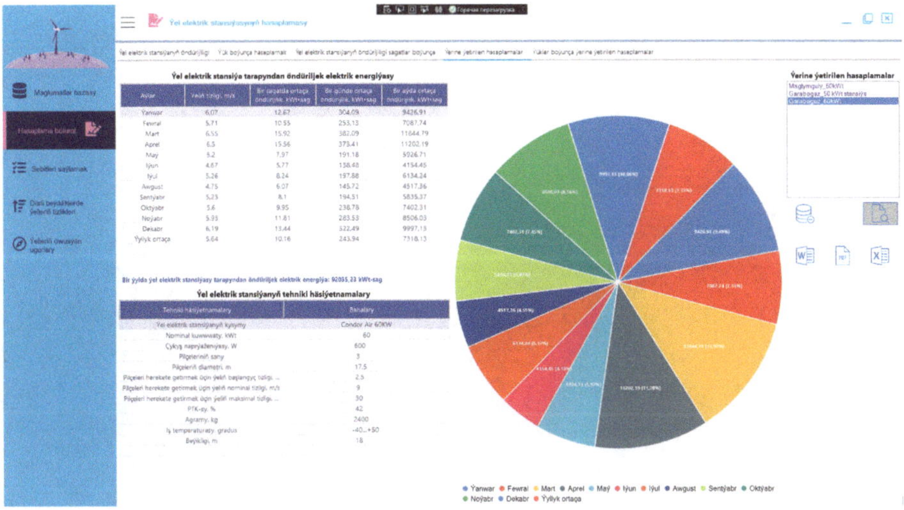

Figure 1.11: Window for performed calculations.

1.4.5 Calculations performed on loads

In this section, it is taken into account that the results of calculations on loads are accessed and stored in the database. Then the calculations are printed or exported to the data storage devices in *pdf*, *doc* and *xls* file formats (Figures 1.12 and 1.13).

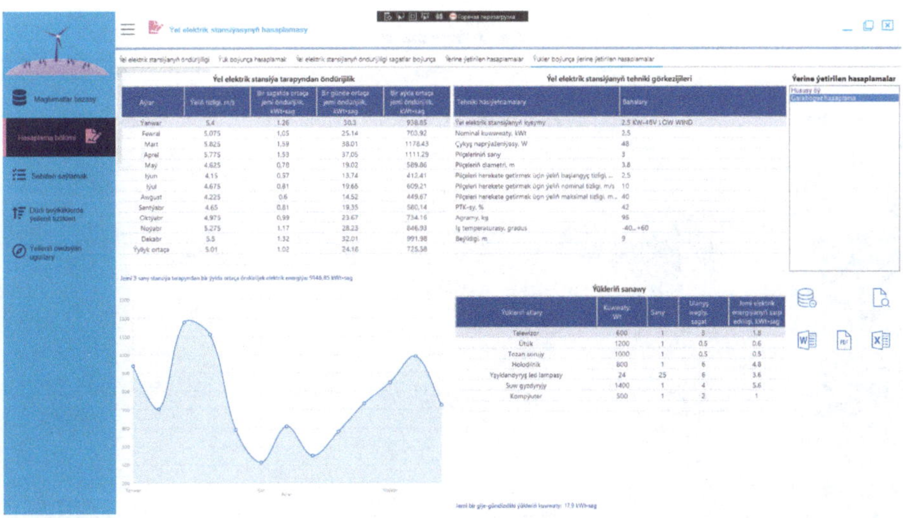

Figure 1.12: Window for calculation performed on loads.

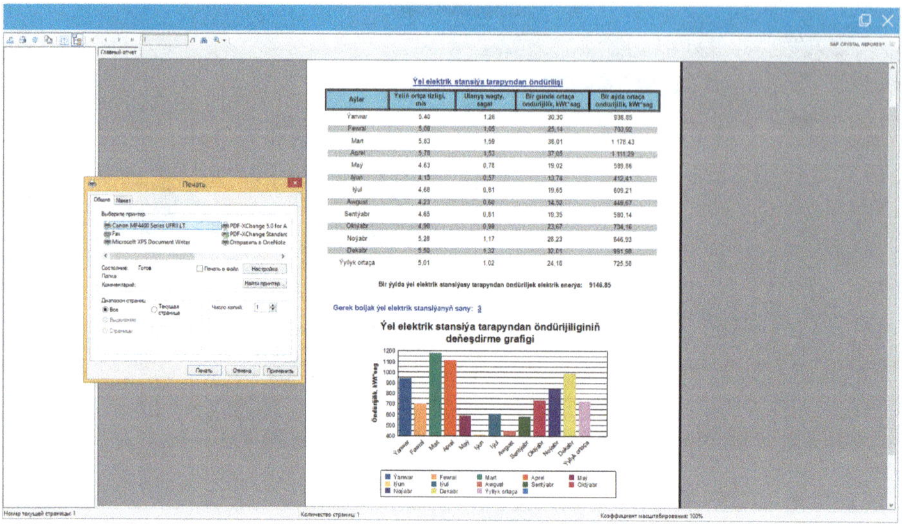

Figure 1.13: Window for printing out the performed calculation.

1.4.6 Area selection

In the application program, it is possible to assess the wind energy resources of areas inserted into the database to give analysis of suitable places for installing the wind power station and to select the most suitable areas (Figure 1.14).

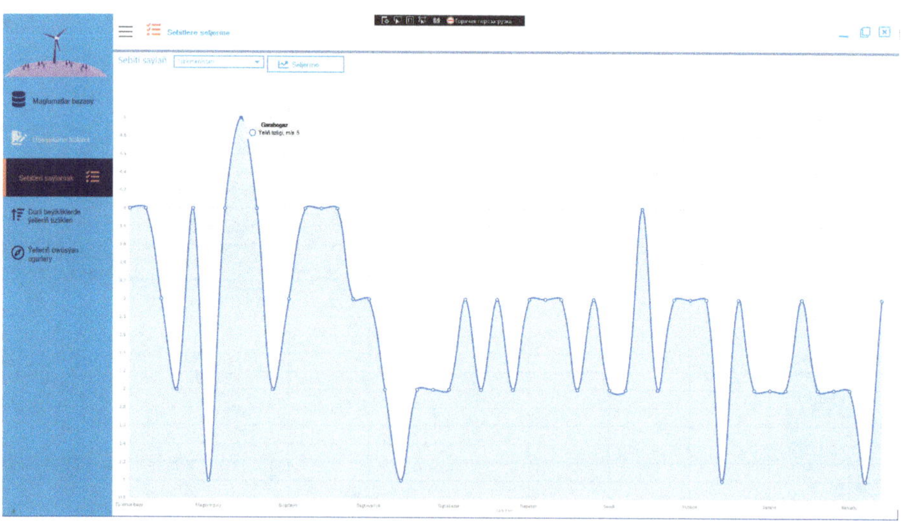

Figure 1.14: Operation window for the selection of area.

1.4.7 Wind speed in various heights

The application program gives an opportunity to determine the wind speed in various heights (between 10 and 100 m). The wind speed at various heights is determined by selecting the most suitable areas (Figure 1.15).

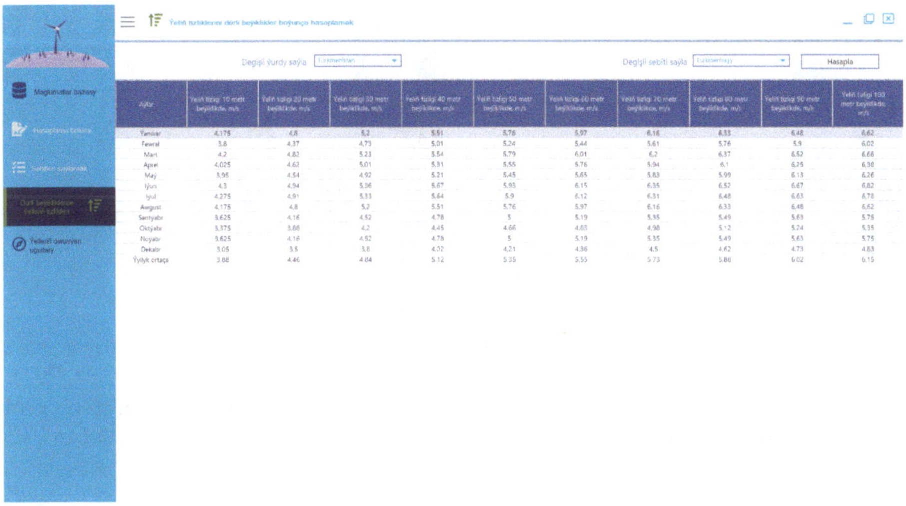

Figure 1.15: Window wind speed determination at different heights.

1.4.8 Wind blowing directions

One of the major indicators of a wind cadaster is the blowing direction of winds. In this section, it is available to define the blowing direction of winds according to the areas existed in the database (Figure 1.16).

1.5 Motivation or future research

This software allows you to quickly carry out projects related to the design and selection of locations for installation, as well as the selection of equipment from various companies.

You can also quickly and easily perform a calculation and, based on the result, see the potential for generating electrical energy. In the future, this program will be able to perform calculations for any country taking into account the entered coordinates.

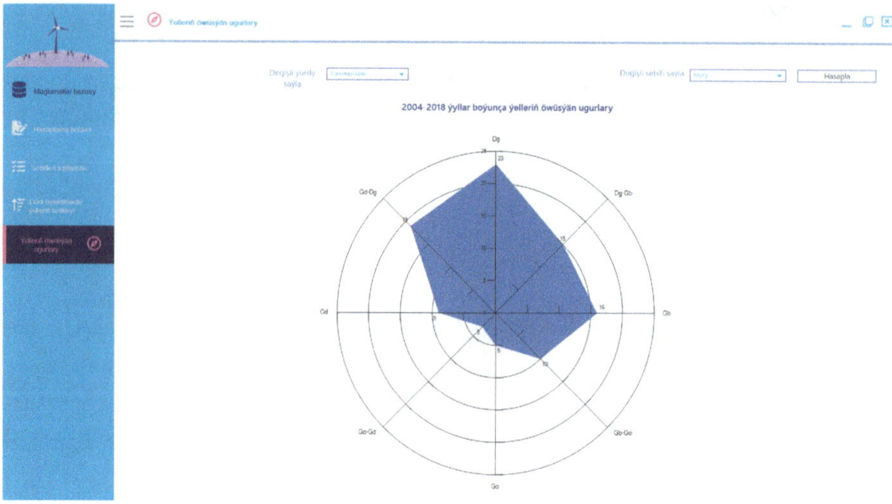

Figure 1.16: Window of wind blowing directions.

I also want to discuss this program at the fifth International Conference on Intelligent Computing and Optimization 2022.

1.6 Conclusion

- As a consequence, the calculation method is proposed through this developed program. According to the undertaken calculations, it is possible to use this program widely during the accomplishment of the project of photoelectric solar station. By taking into account the technically complex issue of designing the photoelectric solar station, this program provides the opportunity of carrying out the selection of the elements of station in a simple method is analyzed on scientific basis.
- As a consequence, calculation and analysis methods are proposed through this developed program. According to the undertaken calculations, it is revealed that this program can be widely used during the working of wind power station projects. Information received from calculations of application program can be consumed to assess the wind energy resource in the region, to determine the construction site of wind energy installations and also to establish the digital database system of wind energy resources in our country.

References

[1] A. Awasthi, A. K. Shukla, S. R. Murali Manohar, C. Dondariya, K. N. Shukla, D. Porwal and G. Richhariya. Review on sun tracking technology in solar PV system. Energy Reports, 6:2020, 392–405

[2] S. P. Mohanty, A. Rout, P. K. Patra and S. S. Sahoo. Soft computing techniques for a solar collector using solar radiation data. Energy Procedia, 109:2017, 439–446

[3] P. Hold, S. Erol, G. Reisinger and W. Sihn. Planning and evaluation of digital assistance systems. Procedia Manufacturing, 9:2017, 143–150

[4] A. Gómez-Espinosa, P. D. Lafuente-Ramón, C. Rebollar-Huerta, M. A. Hernández-Maldonado, E. H. Olguín-Callejas, H. Jiménez-Hernández, E. A. Rivas-Araiza and J. Rodríguez-Reséndiz. Design and construction of a didactic 3-DOF Parallel Links Robot Station with a 1-DOF Gripper. Journal of Applied Research and Technology. JART, 12(3):435–443 (Junio 2014).

[5] C. Vasar, M. Babescu and D.-I. Petrescu. Maximum power point control approach for wind generators. In: Soft Computing Applications: Proceedings of the 6th International Workshop.

[6] Official site Government of Canada/access mode: https://www.nrcan.gc.ca/maps-tools-and-publications/tools/modelling-tools/retscreen/7465

[7] Turkmenistan Golden Age A unique “green” energy project 24.01.2022 w: https://turkmenistan.gov.tm/en/post/60380/unique-green-energy-project

[8] Confirmed by decree of the President of Turkmenistan «energy Saving State Plan for 2018-2014 Years. Ashgabat, 2018y.

[9] “Altyn asyr” Conception of Turkmen lake areas in 2019-2025. Ashgabat -2019

[10] Р. Мартин and М. Мартин., “Принципы, паттерны и методики гибкой разработки на языке C#”. Санкт-Петербург – Москва 2011.

[11] Э. Стиллмен. and Д. Грин. “Изучаем C#”.3-е издание. Москва – Санкт-Петербург – Нижний Новгород – Воронеж Ростов-на-Дону – Екатеринбург – Самара – Новосибирск Киев – Харьков – Минск 2014.

[12] Джон ШАРП. “Microsoft Visual C# ПОДРОБНОЕ РУКОВОДСТВО”.8-е издание. Санкт-Петербург • Москва • Екатеринбург • Воронеж Нижний

K. Saryyev

Chapter 2
Potential wind energy in Turkmenistan

Abstract: While Asia as a continent has enjoyed nearly 40% of the total installed wind energy capacity, the contribution of some countries in the region is less significant. Turkmenistan as an important oil and gas producing country is a major exporter of electricity in Central Asia. The country has an enormous potential for wind and solar energy development overshadowed by its wealth of oil and gas.

When choosing a region for the designing of wind installations, it is imperative to take special attention to the nature of the wind. This mainly refers to high spatial and temporal irregularities, as well as wind variability.

With the correct determination of the wind energy potential, it is possible to solve the energy, economic, environmental and social issues of the country. Moreover, there is an opportunity to mitigate climate change on wind power plants basis, and their resources, environmental benefits, goals and objectives on the scientific methodological basis in the field of wind power for the implementation of state programs on energy saving in Turkmenistan and energy supply of the region. On the basis of the performed scientific work, wind energy resources and environmental potentials in the territory of Turkmenistan have been determined, and a database has been created for the compilation of a wind energy cadaster.

Keywords: Wind energy, direction of the wind, graphics of average values of wind speed

2.1 Introduction

In all regions of Turkmenistan, light and medium winds with speed of 0–5 m/s are dominated, and their frequencies are 75–80% of all potential speeds. In coastal areas of Caspian Sea during day time in the summer and in areas of north Kopet Dagh during day and night in the winter, the wind speed at a height of 150 m is 6–9 per second. Areas with winds more than 5 m speed during the year occupy 15% of the total territory of Turkmenistan, and in areas with more than 20% (for modern facilities with a capacity of 250 kW) capacity ratio occupy more than 20% of its territory. Generally, the power capacity of wind in Turkmenistan is estimated to be 640 billion kW/year.

K. Saryyev, Bayramhan Street 62, Mary, Turkmenistan, e-mail: kakageldisaryyev@gmail.com

https://doi.org/10.1515/9783110717655-002

The approximate cost of electricity generated by wind and by gas is comparable. The main hampering factors in development of wind energy in Turkmenistan are unstable and forecast evaluation of energy generation, necessity of additional energy generation sources and accumulating systems.

The use of wind energy is important for remote locations from provision of the central electric energy system.

2.2 Materials and methods

The power energy sector of Turkmenistan is the fundamental sector for national economy. It is the main basis for the working of other sectors in the country. So it is clearly evident that the energy sector has great advantages. Nowadays, the energy sector not only meets the demands of internal consumers but also exports the power energy to neighbor countries in accordance with the established procedure on contract basis.

Nevertheless, one of the prior tasks of the international society is to widen the usage of renewable energy sources, since it is connected with issues such as prevention of climate change and environmental damage. Significant ways are being implemented to solve these issues effectively in Turkmenistan.

To diversify the fuel and energy means of Turkmenistan, expand the export capacity of natural gas and electric energy, provide remote regions with favorable and clean energy, improve people's well-being, develop the industry and reach the purposes of Paris convention about sustainable development and climate change, precise issues are raised in the implementation of "The National Strategy on development of renewable energy until 2030 in Turkmenistan" program. This was adopted by the president of Turkmenistan according to the decree no. 2007 dated December 4, 2020: "Studying and preparing offers about suitable places and areas to install the industrial solar and wind power stations." The framework of implementation of "State Program of energy saving for 2018–2024 years" was in accordance with the decree no. 674 by the president of Turkmenistan on February 21, 2018: "Drafting the proposals on selection of locations for installing solar, wind energy complexes and also solar and wind cadasters in the territory of Turkmenistan and assessment of solar and wind energy sources between the years of 2018–2021."

Obviously, the fair weather and geographical conditions of Turkmenistan create a vast opportunity to develop renewable energy sources in our country. These opportunities will assist in improving the energy supply of residential areas remote from the central power supply system and in reducing the exhaustion of harmful gases to the environment. Moreover, using solar and wind energy in our country will increase the capacity of "green economy."

About 40% of the region of Turkmenistan is considered suitable for using wind energy, especially in the north-west territory of our country. In this region, an average wind speed is higher than 4 m/s in a year. The specific capacity of wind stream is high in the north coasts of Caspian Sea and its degree is equal to 110–135 W/m^2. In the whole region of Turkmenistan, there is a predominance of light and moderate winds with speed from 0 to 5 m/s. There is a predominance of winds with speed of 6–9 m/s only during daytime on the coastline of Caspian Sea, both day and night during summer, and 1,500 m above heights of north slopes of Kopet Dagh Mountain in winter. In Turkmenistan, 15% of its region has the annual average wind speed of more than 5 m/s and in 20% of its region, the usage coefficient of capacity (for modern facilities with capacity higher than 250 kW) is more than 20%. Generally, it can be assessed that the wind energy capacity is 640 billion kW annually in Turkmenistan [1].

In general, the global wind energy production is 59 GW in 2018. As a result, the total fixed capacity of wind energy facilities was increased to approximately 11% varying from statistics at the beginning of 2018 and was approximately 622 GW. In the wind energy sector, it is expected to decrease the cost of power energy for land and sea facilities by more than 50% (0.054 USD/ h) from 2018 to 2030 [2].

On the basis of these statistics, using the technology in the conversion of wind energy to power energy is economically feasible. The values can be fundamentally determined in complete consumption capacities which are being produced. To overcome this issue effectively, it is necessary to define the precise specifications of wind energy resources, which are located in the regions of consumers, who will be supplied with power energy. Therefore, if the instability of wind energy characteristics in the determined geographic latitude and various placements of generated energy sources under different circumstances are taken into account, then it raises the vitality of undertaking the project works initially on the basis of precise assessment of energy possibilities of wind energy [3].

To calculate the locations for designing the wind energy stations, the precision of wide information about wind conditions of areas, natural process and conversion of wind energy to power energy stands as a significant challenge. It is not enough to obtain information only from general meteorological stations, but it is recommended to benefit information obtained from the NASA satellite database and meteorological stations in the areas of installment of station at various heights at least during a year and to contrast these information. Getting such characteristics is one of the major problems of wind energy cadaster. In the research paper, the calculation method for selecting the locations to place the wind energy structures for regions of Turkmenistan and evaluating the wind energy resources is proposed.

The research paper used the precise values of respective measurements obtained for determining the solar radiation level, water surface evaporation, atmospheric temperature and humidity, the amount and rate of atmospheric precipitation, wind speed and direction in the meteorological station with model "Davis Vantage Pro 2 Plus" installed in the independent energy supply testing laboratory of State Energy Institute of Turkmenistan. The received information is consumed for determining the productivity of wind power installation with 2 kW capacity installed in the institute.

As a consequence, the annual average production capacity of wind power installation is determined under the respective calculations. Additionally, the technical specifications of wind energy installation with 2 kW capacity, which is installed in the institute, are consumed for evaluating the wind energy resources and determining the placement of wind power installations for regions of Turkmenistan.

To determine the wind energy intensity and effective usage, the major characteristic of wind is its average speed in fixed time (in a day and night, monthly and annually).

Then, the annual average wind speed can be defined by the following formula [4–7]:

$$V_{\mathrm{aver.ann}} = \frac{1}{12}\sum_{i=1}^{12} V_{\mathrm{m}} \tag{2.1}$$

where V_{m} is the monthly average wind speed in m/s.

An average value of wind speed received from Mary Meteorology Station between 2015 and 2020 is given in Table 2.1.

Table 2.1: An average value of wind speed received from Mary Meteorology Station between 2015 and 2020.

Years	Months (m/s)												Annual average (m/s)
	I	II	III	IV	V	VI	VII	VIII	IX	X	XI	XII	
2015	3.2	3.7	3.3	3.6	3.4	3.0	4.1	3.2	2.8	3.2	3.3	3.1	3.3
2016	3.3	3.5	3.6	3.4	3.5	4.0	3.6	3.4	3.1	2.6	2.8	3.4	3.4
2017	3.1	3.5	3.6	4.0	3.4	3.9	3.9	3.2	2.7	3.0	2.7	3.1	3.3
2018	3.4	3.5	3.9	3.8	3.5	3.5	3.6	3.2	2.6	2.5	2.9	3.3	3.3
2019	3.8	3.3	3.6	3.7	3.4	2.9	3.5	3.5	3.1	2.0	2.4	2.7	3.2
2020	3.3	3.5	3.7	3.9	3.3	3.0	3.8	3.3	3.0	2.9	2.6	3.2	3.3

One of the major symptoms of wind energy is its energy cadaster. It is a combination of aerological and energy features of wind and allows to determine its energy value and possible operation schedule.

Major features of wind energy cadaster are as follows:

- Annual average speed of wind
- Annual and daily direction of wind
- Types and parameters of functions of recurrence and distribution of its speed
- Maximum wind speed
- Distribution of wind in its periods and energy losses
- Definite power and wind energy
- Wind energy sources of area

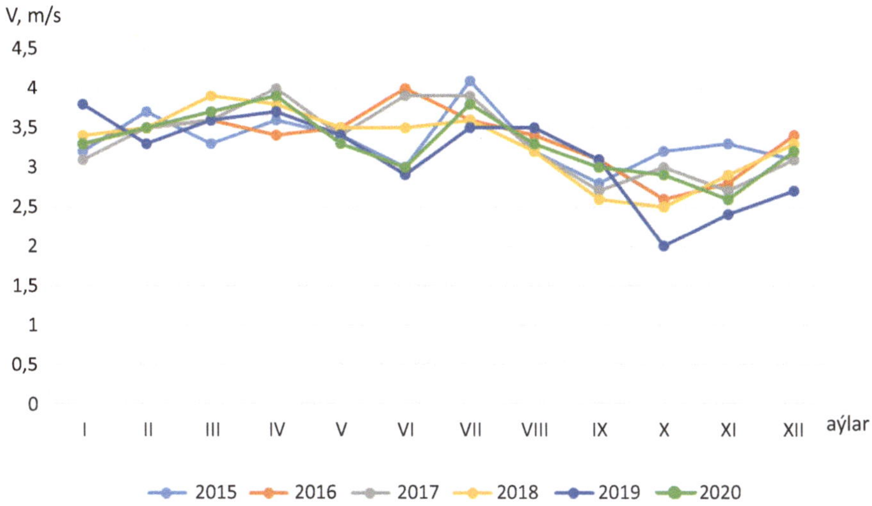

Figure 2.1: Comparison graphics of average values of wind speed in Mary during 2015–2020.

Figure 2.1 has graphically illustrated the Recurrence of wind speed is one of the major features of cadaster characteristics. It shows the duration of wind blowing in the same or different speed during the experiment. By using this characteristic, the wind energy source is discovered, and the major energy features that reveal the efficiency and possibility of consuming wind energy are determined.

Recurrence of wind in its blowing directions in accordance with Mary Meteorology Station during 2004–2018 (Table 2.2).

An average capacity (P_{average}) of wind flow in 1 m^2 area of Mary can be defined by the following formula:

$$P_{\text{average}} = \frac{(\rho \cdot S \cdot V^3_{\text{average}})}{2} \tag{2.2}$$

$$= \frac{(1.225 \cdot 1 \cdot (3.4)^3)}{2} = 22.01 \text{ W}$$

where ρ is the air density (in usual conditions 1,225 kg/m^3); S is the area of cutting of wind flow (surface area that existed after rotation of wind power station blades); V^3_{average} is the annual average wind speed.

An average power energy generated by wind power station with 2 kW capacity is related to the surface area formed by circulation of station blades and wind speed, which is equal to the following value under our conditions [7–11]:

$$S = \frac{\pi \cdot d^2}{4} S = \frac{\pi \cdot d^2}{4} = \frac{3.14 \cdot 3.3^2}{4} = 8.55 \text{ m}^2 \tag{2.3}$$

where d is the diameter of wind power station blades, m.

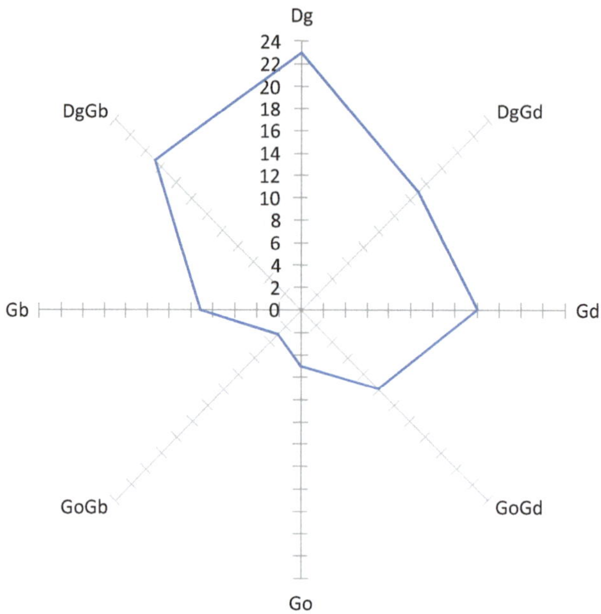

Figure 2.2: Directions of wind blowing during 2004–2018.

Table 2.2: Recurrence of wind in its blowing directions in accordance with Mary Meteorology Station during 2004–2018.

Wind blowing directions	%
North	23
Northeast	15
East	16
Southeast	10
South	3
Southwest	3
West	9
Northwest	19

Figure 2.2 has shown the average capacity of wind flow (P_{average}) can be defined by the following formula in the surface area formed by the circulation of blades of wind power station with 2 kW capacity:

$$P_{\text{average}} = \frac{(\rho \cdot S \cdot V_{\text{average}}^3)}{2} = \frac{(1.225 \cdot 8.55 \cdot (3.3)^3)}{2} = 188.2\ \text{W} \tag{2.4}$$

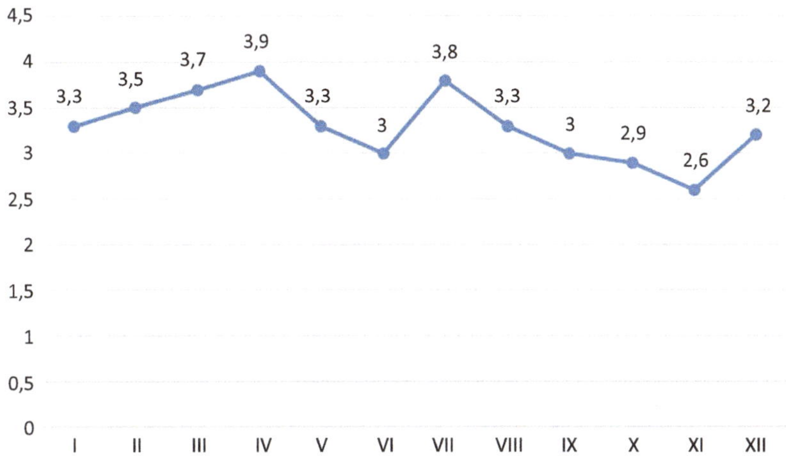

Figure 2.3: An average value of annual wind speed in Mary city.

Figure 2.3 captured the annual production of power energy ($W_{\text{aver.ann.}}$) of the wind power station will be equal to the following value:

$$W_{\text{aver.ann.}} = \frac{(24 \cdot 365 \cdot P_{\text{aver}})}{1{,}000}$$

$$= \frac{(24 \cdot 365 \cdot 188.2)}{1{,}000} = 1{,}648.63 \text{ kW} \cdot \text{h/year} \tag{2.5}$$

If we refer to the internet data, windy days in 2019 are 326 days and windless days are 39 days (https://rp5.ru website information) [12–14].

Table 2.3: Technical characteristics of wind power station with 2 kW capacity.

S. no.	Name of the technical characteristics	Value and unit
1	Nominal capacity	2,000 Wt
2	Nominal voltage	96 W
3	Diameter of blades	3.3 m
4	Initial speed of rotation of blades	3 m/s
5	Nominal speed of rotation of blades	10 m/s
6	Limited speed of rotation of blades	45 m/s
7	Device weight	48 kg
8	Number of blades	3
9	Operation temperature	From −40 to +80 °C

Based on the information received from Table 2.3 conducted calculations, wind energy resources for Mary city were evaluated and an average value of annual wind speed is determined. These values are considered as average value ploted in Figure 2.4 to determine the installaiton places of wind power stations in the region.

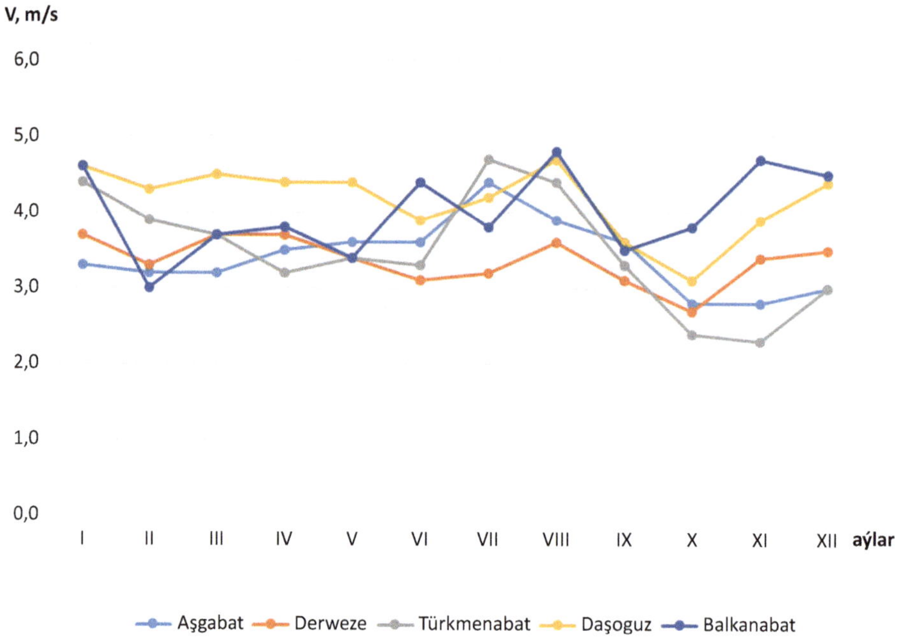

Figure 2.4: An average value of annual wind speed for regions of Turkmenistan.

2.3 Results and discussion

The received information from the research work is useful to assess the wind energy resources for regions of Turkmenistan and clarify the installation places of wind power stations.

Average values of wind speed obtained from meteorology stations in regions of Turkmenistan between 2004 and 2018 are given in Table 2.4.

Table 2.4: Average values of wind speed obtained from meteorology stations in regions of Turkmenistan between 2004 and 2018.

S. no.	Regions of Turkmenistan	Location of meteorological stations in the regional territory	Geographical latitudes (degrees)		Average wind speed (m/s)	Annual average power energy productivity of wind power station with 2 kW capacity (kW·h/ann.)
			North latitude	East longitude		
1.	Ashgabat	Ashgabat	37.9	58,3	3,4	1,803.07
2.	Ahal	Derweze	40.10	58.24	3.4	1,803.07
3.	Mary	Mary	37.6	61.8	3.3	1,648.63
4.	Lebap	Türkmenabat	39.1	63.6	3.8	2,517.25
5.	Dashoguz	Dashoguz	41.8	59.8	4.4	3,907.82
6.	Balkan	Balkanabat	39.5	54.4	4.4	3,907.82

2.4 Conclusion

1. It is found out that information received as a result of conducted calculations can be used to assess the wind energy resources for that area and also regions of Turkmenistan and to determine the installation places of wind power stations.
2. By using the technical characteristics of wind energy installation with 2 kW capacity, the annual average amount of power energy, which is generated by various capacity wind power stations, is calculated.
3. On the basis of information received from meteorological stations in the regions, a comparison graph of average values of wind speed and a graph of directions of wind flow are drawn.
4. Annual average values of wind speed are determined for Mary city.

References

[1] State program approved by the Decree of the President of Turkmenistan, A.:2018, 2019.

[2] A. G. Babaev. Physical geography of Turkmenistan, Toolkit A., 2014.

[3] A. Dzhumayev and H. Sultanov. Fundamentals of energy saving. Science, Ashgabat, 2018.

[4] Key word energy statistics, 2009. International Energy Agency.

[5] V. M. Kargiev, et al. Wind power engineering guidelines for the use of wind turbines of small and medium power. M.: 2001.

[6] https://rp5.ruwebsiteinformation.

[7] S. S. Dorjiev, E. G. Bazarova, S. V. Pimenov and M. I. Rozenblum. Development of wind power installations with the accelerator of an air stream for areas with a low speed of wind. Journal of Physics: Conference Series,1111, 012053, 2018.

[8] Wind Energy Projects in Cold Climate (Edition 2011). Submitted to the Executive Committee of the International Energy Agency Programme for Research, Development and Deployment on Wind

Energy Conversion Systems. 2012. Available at: http://ieawind.org/index_page_postings/June7posts/task19coldclimaterpapproved05.12.pdf. Accessed to: 12 Apr 2019.

[9] A. R. Erma, P. L. Kamani and R. R. Kapadia A review on grid power quality improvement in wind energy system using STATCOM with BESS. International Journal of Emerging Technologies and Innovative Research, 2(1):2015.

[10] IRENA. Renewable Capacity Statistics. 2016. Available at: https://www.irena.org//media/Files/IRENA/Agency/Publication/2016/IRENA RE Capacity Statistics 016.pdf. Accessed to: 17 Apr 2019.

[11] J. O'Sullivan Two utilities opt out of cape wind., January 2015. Available at: http://www.bostonglobe.com/metro/2015/01/06/major-setback-for-cape-wind. Accessed to: 06 Apr. 2019.

[12] E. V. Nasyrova, N. F. Timerbaev, O. V. Leukhina and I. Y. Mazarov. Analysis of wind monitoring data in the Republic of Tatarstan" Scientific and practical peer-reviewed journal "Izvestiya Vysshikh educational institutions. ENERGY PROBLEMS "Izvestia vysshikh uchebnyh zavedeni". 21(6):39–50, 2019.

[13] S. N. Salikeeva and F. T. Galeeva. Review of methods for obtaining alternative energy. Science Journal "Vestnic Kazan Technology University, 15(8):57–59, 2012.

[14] E. V. Kuznetsova and P. A. Puganova. Economic feasibility of using wind energy. Science, Education, Innovations, 14–16, 2016.

[15] Impact of Wind Power Generation in Ireland on the Operation of Conventional Plant and the Economic Implications. 2004. Available at: https://docs.wind-watch.org/EirGrid-WindImpact-Main.pdf.: Accessed to:18 May 2019.

M. Orazberdiyeva

Chapter 3
Potential of using biogas technology in Turkmenistan

Abstract: This chapter involves the development of independent and waste-free renewable energy technologies using renewable energy sources. Therefore, it is possible to prevent environmental pollution. The development and introduction of environmentally friendly and economically viable technologies is one of the key tasks in this area.

This chapter examines the possibilities of renewable energy sources in Turkmenistan and their current state. Research has also been carried out on the achievement of ecological well-being in the country.

The importance of producing biogas from agricultural and agroindustrial waste is very high. At the biogas installation, biogas is formed as a result of the fermentation of the mixtures. The remaining waste serves as a high-quality biofertilizer for agricultural crops. Taking into account the advantages of biogas utilization using renewable energy sources, research has been carried out in the experimental field of the Center for Renewable Energy Sources of the State Energy Institute of Turkmenistan. Experiments on the production of biogas by oxidation of manure waste using an airless (anaerobic) method have been carried out in a specially prepared small-scale biogas installation.

According to the calculations, the amount of biomass pumped into the biogas plant was 438.5 m^3, the daily output of biogas was 548 m^3 and the amount of biogas consumed by the livestock complex for heat and electricity was 147.95 m^3/day. The remaining amount of biogas, i.e., 548–147.95 = 400.05 m^3, is recommended for use by other consumers or for the needs of the population.

As can be seen from the above calculations, the introduction of biogas technology in livestock complexes is leading to be environmentally friendly and economically profitable. This testifies the fact that the introduction of biogas technology in livestock farms will be convenient in all respects.

Keywords: Renewable energy sources, achieving ecological well-being and increasing economic efficiency through the extraction of biogas from waste, the introduction of biogas technology in livestock complexes

M. Orazberdiyeva, Bayramhan Street 62, Mary, Turkmenistan, e-mail: mahribanorazberdiyeva@gmail.com

https://doi.org/10.1515/9783110717655-003

3.1 Introduction

Turkmenistan is known as a powerful country in terms of energy and is one of the main energy exporters in the region. Large-scale energy projects are being successfully implemented in our country; power transmission lines and electricity distribution stations are being built; and all corners of our capital and the country are being provided with electricity. As a result of the large-scale activities of our state on a global scale, the international prestige of our motherland is increasing day by day.

Today, the production of electricity and heat energy from renewable energy sources is environmentally friendly and economically viable. Along with the increase in living standards, production and industrial enterprises are also increasing in number. Electricity and heat energy are required for their continuous, uninterrupted operation. This leads to ecologically unfavorable and economically high costs when conventional methods are used.

The most effective way to solve these issues is to develop independent, waste-free technologies that generate electricity and heat by using renewable energy sources. As a result, it is possible to prevent the environmental pollution. The development and introduction of environmentally friendly and economically viable technologies are among the key tasks in this regard.

Today, the goal of adopting renewable energy globally is to reduce air pollution and, as a result, prevent climate change, produce economically viable products, prevent environmental pollution and reduce unpleasant odors with the adoption of waste-free technologies due to waste recycling.

One of the technologies that can work autonomously (independently) without waste by using renewable energy sources is the extraction of biogas from organic wastes, such as agricultural livestock and poultry waste, plant waste, and waste from agricultural enterprises. Here, biogas is formed as a result of airless (anaerobic) oxidation of organic waste.

Energy efficiency and renewable energy are reliable and affordable ways to ensure a significant reduction in emissions. The combination of these two elements can reduce CO_2 emissions associated with energy production by more than 90% using reliable, cost-effective and widely used technologies.

The introduction of new innovative technologies in the field of renewable energy sources allows us to move beyond conventional energy sources. Renewable energy has many different types of sources, one of which is biomass energy.

Biomass energy: Biomass includes wood and wood waste, agricultural waste such as manure and plant waste, especially grown agricultural crops for biomass, and organic and solid waste from production and industrial enterprises [1].

It can be used to generate energy by burning biomass or by using various technological processes (e.g., fermentation) to produce biogas or liquid biofuels. These, in turn, can be converted into electricity and heat (e.g., by burning or creating fuel cells) or can be used as fuel for road vehicles. The most common types of fuels are ethanol, biogas and biodiesel.

The use of biomass as fuel in biogas production is a significant new phenomenon in energy and resource conservation, based on agricultural and industrial waste [1, 2].

Energy production from biogas is mainly based on its use in biomass-rich areas with agricultural or agroindustrial waste. The current production chain includes pretreatment (disintegration), acidification (fermentation), biogas extraction and its energy conversion from the use of wastes [3].

3.2 Literature review

The specifics of the use of biogas were known before our era. Scientific research on biogas and its properties began only in the eighteenth century. In 1764, Benjamin Franklin described a test that succeeded in setting fire to a swampy lake. European scientists took the first steps in studying the practical application of biogas after discovering the presence of methane in swamp gas and finding its chemical formula.

In 1875, the Russian scientist S.V. Popov studied the effect of temperature on the amount of gas. He found that even at 6 °C. River sediments begin to emit biogas, and as the temperature rises, its volume increases [4].

The first regular biogas research was conducted by Italians. Alessandro Volta also conducted a study on electricity and named the voltage measurement unit as “volt.” He extracted swamp gas from lake sediments in northern Italy. This gas combustion has been tested several times.

The British physicist Faraday experimented with swamp gas and studied it. It was not until 1821 that the explorer Avogadro was able to determine the chemical formula of methane (CH_4). The famous French bacteriologist Pasteur experimented with biogas in 1884 and removed it from extreme fertilization. He first proposed the use of fertilizer in stables for production, and the biogas obtained was used for street lighting [5].

In the field of biogas extraction from wood waste, it is possible to heat houses, industrial buildings and baths, as well as dry grains and fruits with the help of a heat generator [6].

Mao et al. [7] provided a comprehensive overview of research advances in anaerobic fermentation for biogas production. The review includes a discussion of factors affecting efficiency (temperature, pH, C/N ratio, OLR and retention time), accelerators (green biomass, organic culture and inorganic additives), reactors (conventional anaerobic reactors, sludge retention reactors and anaerobic membrane reactors) and biogas process of anaerobic digestion (lignocellulose waste, municipal solid waste, food waste, livestock manure and activated sludge waste) based on the characteristics of the substrate and discusses the application of each of the above aspects.

Factors affecting efficiency are critical for anaerobic fermentation, as they play an important role in biogas production and determine the metabolic conditions for microbial growth. As an additive, the accelerator is seen not only as a source of nutrients but also as a means to improve biodegradability.

Angelidaki et al. [8] presented a critical review that summarizes modern technologies for improving the quality of biogas with special attention to the emerging processes of biological methane formation. The review includes a comprehensive description of the basic principles of methods for improving the quality of biogas, as well as scientific and technical results related to the efficiency of biomethane production.

Al Seadi et al. [9] provided the main characteristics of biogas feedstock, including their suitability, availability, fermentation and purity. Although biogas feedstocks can be classified according to their criteria, in this chapter, they are grouped according to their sectors of origin (agriculture, industry, municipal and water industry), highlighting the advantages and disadvantages (limitations) of the most commonly used raw materials in each category. It also emphasizes the fundamentals of availability, supply and logistics of raw materials and their impact on issues related to the location of a biogas installation.

In IEA Bioenergy [10, 11], efforts are made to reduce the impact on equipment that uses biogas because biogas contains not only methane, carbon dioxide and water, but also other compounds in small quantities. These impurities can cause corrosion and mechanical wear, and can also lead to unwanted emissions.

Sun et al. [12] systematically reviewed the current state of biogas purification and modernization technologies, including product purity and the presence of impurities, methane recovery and loss, efficiency improvements, as well as investment and operating costs.

In addition, the authors carefully studied the possibilities of using biogas on an industrial scale and the corresponding requirements for gas quality. Based on the comparison of the technical characteristics of modernization technologies, the specific requirements for the different uses of biogas, and the corresponding investment and operating costs, recommendations are made for the appropriate technology.

Rafiee et al. [13] and Mezmur and Bogale [14] provided a general definition of the composition of biogas and its types, as well as its purification. Since all types of biogas must be purified and upgraded before being delivered to consumers, the key task of the biogas supply chain in practice is to purify and improve the quality of the biogas for the consumer.

3.3 Materials and methods

Biogas is a gaseous fuel. Its main components are methane (CH_4, 55–70%), carbon dioxide (CO_2, 28–43%) and a small amount of hydrogen (H2O). When 70% of 1 kg of organic matter is biologically decomposed, it produces 0.18 kg of methane, 0.32 kg of carbon dioxide, 0.2 kg of hydrogen and 0.3 kg of waste. One cubic meter of biogas can generate 2.4 kW·h of electric energy and 3.2 kJ·h of thermal energy [15].

To extract biogas from organic matter and waste (agricultural waste, organic fertilizers from farm animals and birds, plant remains and others) is very important. Using different wastes to extract biogas leads to the reduction of environmental pollution, greenhouse gases and unpleasant odors, which protects our environment. Extracting biogas fuel using cheap raw materials to get high-quality biogas fuel is very profitable. High-quality organic fertilizers obtained from raw organic fertilizers and excreta of farm animals are environmental pollutants. As agricultural plants absorb raw organic fertilizers from animals quickly, underground waters and the air are polluted. This leads to ground infection with hazardous microorganisms. The viability of bacteria is endless, and seeds of weeds do not lose their properties.

Biofertilizers processed in a biogas unit are high-quality fertilizers for plants. During anaerobic fermentation in a biogas unit, nitrogen, phosphorus and potassium are converted into a form readily available for plants. Dermatitis-causing agents in animals and weed seeds become unviable.

The importance of biofertilizers in agriculture is as follows:

- Organic remains and organic fertilizers of animals contain organic matters (NPK minerals).
- In a biogas unit, organic materials are converted into methane and CO_2.
- Nitrogen, phosphorus and potassium minerals (NPK) from an organic state are directly converted into an available state for plants.
- Effectiveness of nitrate increases from 60% to 85–90%.
- Pouring the remains of the fermentation process onto the land improves the structure of the soil.
- Improves the activity of underground phosphorus.
- Activates the formation process of nitrate in the ground. Stimulates the development of common nitrogen proteins and increases the extraction of carbon dioxide from the ground.
- Intensifies the circulation of phosphorus as well as amide and hydrogen nitride in a plant.

Biofertilizers improve the physical and mechanical properties of the soil. Soil circulation processes are strengthened as well as their biological activity. They have a significant impact on harvest capacity: autumn wheat by 15–20%, sugar beet by 20%, corn (maize) by up to 15–20% and potatoes by up to 30%.

Superiority of biofertilizers other organic fertilizers is as follows:

- **Absence of weed seeds:** Pig and cattle manure and peat contain large amounts of weed seeds. On average, 1 ton of fertilizer manure contains about 10,000 different weed seeds. These seeds retain their viability even after passing through the animals' stomachs and intestinal tracts. This can lead to a loss of 57 hundredweight of harvest per hectare.

Table 3.1: Composition of biofertilizer after processing in a biogas unit with solid part having 75% humidity.

t/b	Biofertilizer (mass after fermentation)	N	Chemical composition (kg/ton)			MgO
			NH_4N	P_2O_5	K_2O	
1	Pig fertilizer manure	5.9–6.5	1.4–2.0	5.3–5.8	6.1–6.3	1.5–1.8
2	Cow fertilizer manure	4.3–5.0	1.0–1.2	2.7–2.9	7.5–7.8	1.3–1.5
3	Bird fertilizer manure	17–18	3.0–3.5	10–10.9	8.0–8.8	3.5–4.2
4	Verdant grass	3.2–3.5	0.7–1.0	1.37–1.4	4.2–4.5	0.5–0.6
5	Corn silo	3.7–4.0	1.2–1.3	1.3–1.4	4.2–4.5	0.8–1
6	Remains of sugar beet	5.0–6.2	–	3.3–3.5	4.2–4.5	1.2–1.6
7	Bloody remains of animals	10	1.8–2.0	20–25	3.0–3.5	2.5–2.6

Table 3.2: Composition of biofertilizer after processing in the biogas unit with liquid part having 95% humidity.

t/b	Biofertilizer (mass after fermentation)	N	Chemical composition (kg/ton)			MgO
			NH_4N	P_2O_5	K_2O	
1	Pig fertilizer manure	3.1–3.8	1.4–2.0	2.3–2.4	2.1–2.4	0.5–0.8
2	Cow fertilizer manure	1.8–2.2	1.0–1.2	0.8–1.6	2.2–2.8	0.4–0.5
3	Bird fertilizer manure	7.1–8.2	3.0–3.5	6.8–7.9	5.0–5.6	1.5–2.2
4	Verdant grass silo	2.2–2.8	0.9–1.5	1.9–2.3	2.0–2.5	0.5–0.7

- **Absence of microorganisms and microflora of pathogenesis of diseases:** In most cases, organic fertilizers can spread plant infections. However, through recycling, biofertilizers become completely free from pathogenic microflora.
- **Presence of active microflora that helps in intensive growth of plants:** Organic wastes used as fertilizers are either free of microflora or contain them in small amounts.
- **Absence of an adaptation period:** Before fertilizing the ground, manure should be prepared for a long time (6–12 months). Useful properties are partially lost, and the remaining ones start to affect the fertilizing ground after 2–4 years. In contrast, biofertilizers start to affect immediately.
- **Stability in washing off useful elements from the soil:** An average of 80% of organic fertilizers is washed off from the ground in one season, whereas for biofertilizers, it is 15%.
- **Accumulation and storage of nitrogen in large amounts:** Due to the composting of organic remains over a long period, up to 50% of nitrogen is lost. Biofertilizer not only retains nitrogen in its full form but also increases the storage of NH_4 by up to 10–15%.

- **Environmental impact on ground:** Unprocessed organic fertilizers damage the ground by polluting underground water. In this case, biofertilizer is considered to be an environmentally clean alternative. Table 3.1 and Table 3.2 has explained the comparison studies on 75% and 95% bioga with solid and humidity.

3.3.1 Superiority of biofertilizer in comparison with mineral fertilizers

- Only 30–35% of mineral fertilizers are processed; the remaining portions stay in products and the ground in nitrate form.
- Nitrate causes swelling in the stomach and intestinal canal. The utilization of nitrate in small amounts over a long period leads to the swelling of thyroid glands. Nitrate functions as a protein reducer and cholesterol increaser in human and animal blood.
- Biofertilizer is able to be processed by itself to up to 100%, resulting in reduced nitrate storage in plants.

Recycling agricultural, industrial and organic underground wastes and remains in modern biogas units reduces pollution of the ground, water and air. Hazardous bacteria and viruses are eliminated, unpleasant odors are reduced and a local energy source is utilized. According to the requirements of environmental protection to use anaerobe technology in recycling of wastes and remains of agricultural and industrial enterprises is very beneficial. The main function of this technology is to prevent the negative impact of wastes and remains on human health and the environment.

Taking into account the advantages of biogas utilization using renewable energy sources, research has been carried out in the experimental field of the Center for Renewable Energy Sources of the State Energy Institute of Turkmenistan. Therefore, experiments on the production of biogas by oxidation of manure waste using an airless (anaerobic) method have been conducted in a specially prepared small-scale biogas installation (Figure 3.1).

In the biogas plant, 40 kg of cattle manure and 20 kg of bird manure were combined, and 150 L of water was added. The mixture (biomass) was stirred every 4 h, and the ambient air temperature was recorded. Biogas began to emit 7–10 days after fermentation. The composition of the obtained biogas was determined, and the results were as follows:

Composition of biogas extracted from the combined fertilizer manure:
Argon – 0.1%
Nitrogen – 3.0%
Hydrogen – 10.3%
Methane – 86.4%

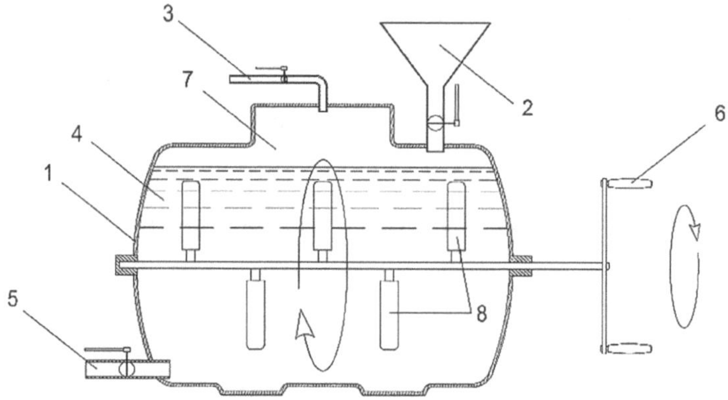

Figure 3.1: Schematic structure of the biogas: 1, bioreactor; 2, funnel; 3, gas outlet; 4, paddle; 5, waste extractor;
6, hand-operated drive unit; 7, biogas accumulation site; 8, mixing paddles.

With the introduction of biogas technology in livestock farms and farms that are far from the central power supply, the problem of heat and energy supply for these farms has been solved. Additionally, when transported over long distances, it will prevent the formation of unpleasant odors in the environment and make manure more accessible for use on crops.

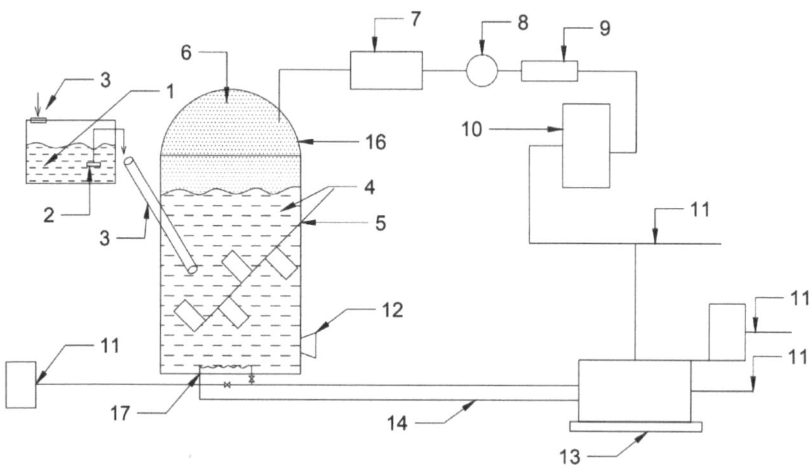

Figure 3.2: Schematic diagram of a biogas plant with a cogeneration system and an accumulator system in livestock complexes: 1, organic waste; 2, suction pump; 3, place of the mixture spill; 4, biofuels; 5, mixing paddle; 6, biogas; 7, low-pressure gas collector; 8, filter; 9, compressor; 10, high-pressure gas collector; 11, consumers; 12, biofertilizer; 13, cogeneration unit; 14, heat carrier for biomass heating; 15, accumulator system; 16, methane tank; 17, heater.

The biogas installation to be placed in the livestock complex (Figure 3.2) is composed of an organic waste disposal plant, a methanizer, a gas collector, a gas purifier filter, a cogeneration plant and a biofertilizer collection unit.

Organic waste is prepared in a facility with special crushing and mixing equipment to form a uniform mixture before the manure is delivered to the methane tank. The mixtures are then fed into the methane tank. Here, the biochemical process of fermentation based on the activity of methane-forming bacteria at an air temperature of 35 °C occurs. As a result, combustible biogas (CH_4: 55–70%; CO_2: 28–43%) is formed.

The biogas is fed to a low-pressure gas collector, which then passes through a gas purifier filter to the high-pressure gas collector. From here, biogas is supplied to consumers in the form of electricity, hot water and biogas. Waste generated in the methane tank is then sent to a biofertilizer facility. From there, it is used for agricultural needs.

A research project on the "Implementation of Biogas Technology in Livestock Complexes" is currently being developed at the Renewable Energy Sources Research and Production Center. The project conducts research to meet the electricity and heat demands of the complex mainly through biogas derived from cattle manure, and to transfer the remaining amount of biogas to consumers.

The biogas plant, which will be housed in the livestock complex, also includes the installation of a cogeneration plant and an energy storage system. It is noted that, according to global experience in similar projects, the deployment of a cogeneration plant in biogas technology has several features. These features include the following:

1. Processing of organic waste
2. Reducing the impact of harmful gases
3. Elimination of costs incurred in technological connections to centralized networks in new construction facilities
4. Continuous maintenance of electricity and heat supply for private production
5. Increase in revenue due to cost savings in electricity and heating
6. Obtaining environmentally friendly and high-quality organic fertilizers
7. Protecting the health of the population by obtaining environmentally friendly products through the use of biofertilizers in agriculture [16]

3.3.2 Project calculation

Calculations for identifying the power of the biogas unit, which will be installed in cattle farm complexes, were carried out according to the average annual consumption rate of raw materials. The power of the unit is calculated indirectly, based on the amount of gas extracted in a year. To satisfy the heat and electric energy demands of the cattle farm complex, cogeneration is incorporated into installation of the project. The remaining gases are planned to be used for domestic purposes or to produce liquid fuel methanol.

If there are about 500 cattle in a cattle farm complex, then each of them produces about 35 kg of biomass with a humidity of 70–80% over 24 h. Therefore, 35 × 500 = 17,500 kg biomass is possible to obtain from them in 24 h.

The amount of valid waste for biogas extraction is [17]

$$B_{\text{b.g.}} = 17{,}500\ \text{kg/day} - \text{night}$$

Identification of the amount of water: To achieve consistent fermentation of biomass, the temperature of water should be 35 °C and added in a ratio of 1:3.

In this case, water consumption over 24 h is as follows:

$$B_{\text{S}} = \frac{B_{\text{b.g}}}{3} = 5{,}833.3\ \text{kg}$$

3.3.3 Identification of the pouring amount of biomass into the bioreactor

Raw materials stay in the bioreactor for 14 days and the fermentation process occurs. The density of the poured raw material [17] is

$$\rho_{\text{b.g.}} = 745\ \text{kg/m}^3$$

Raw material amount during 14 days is

$$M_{\text{b.m}} = 14\left(B_{\text{b.g}} + B_{\text{b}}\right) = 14(17{,}500 + 5{,}833) = 326{,}662\ \text{kg}$$

In this case, the capacity of the bioreactor is as follows:

$$V_{\text{b.m}} = \frac{M_{\text{b.m}}}{\rho_{\text{b.g}}} = \frac{326{,}662}{745} = 438.472\ \text{m}^3$$

Identification of extraction of biogas: As a rule, the gas part of the reactor is 20%, and the active part with biomass is 80% [18, 19]. In this case, the extraction of biogas from $V_{\text{b.m}} = 438.5\ \text{m}^3$ capacity bioreactor will be

$$V_{\text{b.g.}} \frac{438{,}472}{80} \times 100 = 548.09\ \text{m}^3$$

$$V_{\text{b.g.year}} = 360 \cdot V_{\text{b.g}} = 360 \times 548.09 = 197{,}312.04\ \text{m}^3$$

$$B_{\text{h}} = \frac{Q_{\text{tech}} + Q_{\text{ot}} + Q_{\text{BGU}} + Q_{\text{ot}}}{Q_{\text{H}}^{\text{P}} \cdot \eta} = \frac{(50 + 2{,}000 + 900)}{22{,}000 \times 0.95} = 0.141\ \text{m}^3/\text{day} - \text{night}$$

$$B_e = \frac{E_{tech} + E_{bgu}}{Q_H^P \cdot \eta_e^{CG}} = \frac{(40+13)\cdot 10^3}{22000 \cdot 0.35} = 6.883\text{m}^3/\text{day} - \text{night}$$

This is equal to 147,993.7 m^3 of natural gas [20].

3.3.4 Identification of the amount of biogas consumption in cattle farm complexes and for personal needs

The amount of necessary biogas during 24 h to produce electrical energy in cattle farm complexes and for personal needs [21, 22] is calculated according to the following formula:

$$B_{\text{b.g}} = B_{\text{v}} + B_{\text{e}} \tag{3.1}$$

where B_{h} is the consumption of biogas to produce heat:

$$B_{\text{h}} = \frac{Q_{\text{tech}} + Q_{\text{ot}} + Q_{\text{BGU}} + Q_{\text{ot}}}{Q_{\text{H}}^{\text{P}} \cdot \eta}, \ \text{m}^3 \tag{3.2}$$

where Q_{tech} is the heat energy consumed for technological needs of the cattle farm complex, according to Goldstein and Vitart [23]; $Q_{\text{tech}} = 50\text{kW}\cdot\text{h/day} - \text{night}$; Q_{ot} is the consumption of heat energy in the heat system of cattle farm system, according to Struyk and Sposito [24]; $Q_{\text{ot}} = 2{,}000\text{kW}\cdot\text{h/day} - \text{night}$; Q_{BGU} is the heat energy consumption in biogas unit; $Q_{\text{BGU}} = 900\text{kWt}\cdot\text{h/day} - \text{night}$; Q_{H}^{P} is the low hea of biogas combustion; $Q_{\text{H}}^{\text{P}} = 22{,}000\text{kJ}\cdot\text{h/day} - \text{night}$; η_{he} is the heat producer unit efficiency; $\eta_{\text{he}} = 0.95$.

Using formula (3.2)

$$B_{\text{h}} = \frac{Q_{\text{tech}} + Q_{\text{ot}} + Q_{\text{BGU}} + Q_{ot}}{Q_{\text{H}}^{\text{P}} \cdot \eta} = \frac{(50 + 2{,}000 + 900)}{22{,}000 \times 0.95} = 0.141\,\text{m}^3/\text{day} - \text{night}$$

B_{e} is the consumption of biogas to generate electricity [25]:

$$B_{\text{e}} = \frac{E_{\text{tech}} + E_{\text{bgu}}}{Q_{\text{H}}^{\text{P}} \cdot \eta_{\text{e}}^{\text{CG}}} \tag{3.3}$$

where E_{tech} is the electric energy consumption for technological needs, according to Gu and Chen [25]; $E_{\text{tech}} = 40\text{kW}\cdot\text{h/day} - \text{night}$; E_{bgu} is the electric energy consumption of biogas unit; $E_{\text{bgu}} = 13\text{kW}\cdot\text{h/day} - \text{night}$; $\eta_{\text{e}}^{\text{KGU}}$ is the cogeneration unit efficiency; $\eta_{\text{e}}^{\text{KGU}} = 0.35$.

Using formula (3.3)

$$B_{\text{e}} = \frac{E_{\text{tech}} + E_{\text{bgu}}}{Q_{\text{H}}^{\text{P}} \cdot \eta_{\text{e}}^{\text{CG}}} = \frac{(40+30)\times 10^3}{22{,}000 \times 0.35} = 6.83\,\text{m}^3/\text{day} - \text{night}$$

$$B_{bg} = B_t + B_e = 141.15 + 6.8 = 147.95 \text{ m}^3/\text{day} - \text{night}$$

According to calculations, the amount of biomass put into the biogas unit is equal to 438.5 m^3. The production of biogas per day-night is 548 m^3. The amount of biogas consumption required to provide the cattle farm complex with electric and heat energy is equal to 147.95 m^3/per day-night. The remaining 548–147.95 = 400 m^3 amount of biogas is offered to use for other consumers or for population needs.

According to the aforementioned calculations, the introduction of biogas technology into the cattle farm complex is not only environmentally clean but also profitable.

It is evident that the introduction of biogas technology to cattle farmers is profitable in all aspects.

3.4 Results

According to the calculations, the amount of biomass pumped into the biogas plant was 438.5 m^3, the daily output of biogas was 548 m^3 and the amount of biogas consumed by the livestock complex for heat and electricity was 147.95 m^3/day. The remaining amount of biogas, i.e., 548–147.95 = 400.05 m^3, is recommended to be used for other consumers or for the needs of the population.

3.5 Conclusion

The “Renewable Energy Sources” Scientific and Production Center of the State Energy Institute of Turkmenistan is involved in the development of innovative, waste-free self-operating biological complexes. There is development of a combined cycle solar plant project with heating buildings and birdhouses operating on the basis of biogas extracted from waste processing, solar energy and the heat generated by birds’ bodies. The development of technology that uses wastes to extract biogas fuel burning in steam tanks is also underway. The installation of electric generators, birdhouses and biogas units into renewable and large agricultural complexes, which use local types of fuels, is considered to be carried out.

Taking into account the aforementioned examples, biogas production is evidently eco-friendly, profitable and highly valued as a fertilizer in the agricultural field. Using biofertilizers as a plant fertilizer in agriculture plays a great role in obtaining a high-quality and rich harvest.

References

[1] В. Е. Фортов and О. С. Попель. ВОЗОБНОВЛЯЕМЫЕ ИСТОЧНИКИ ЭНЕРГИИ В РОССИИ. Возобновляемая энергетика: проблемы и перспективы. Материалы II Международной конференции. 27–30 сентября 2010г. Под ред. д.т.н. А.Б.Алхасова – Махачкала: ИП Овчинников (АЛЕФ), 2010. 448 с.

[2] В. Тихонравов. С.Т 46. Ресурсосберегающие биотехнологии производства альтернативных видов топлива в животноводстве: науч. аналит. обзор. – М.: ФГБНУ «Росинформагротех», 2011. 52 с.

[3] Леся Матиюк/FNR e.V., Марко Бобцин/FNR e.V., Йоахим Шнур/GFA Consulting Group GmbH. Создание биогазовых установок с целью решить экологические проблемы, вызванные выбросами сельскохозяйственных и агропромышленных отходов и остатков. Продвижение устойчивого производства и использования биоэнергии в Российской Федерации и Украине. www.bio-prom.

[4] Н. А. Чадова and А. Ю. Чадов. Технологии производства биогаза и перспективы его применения в Росии. Студенческий научный форум – 2017 IX Международная научная конференция. Служба Технической Поддержки. support@rae.ru.

[5] Барбара Эдер Хайнц Шульц. Биогазовые установки Практическое пособие. Перевод с немецкого выплонен компанией Zorg Biogas в 2011 г. Под научной редакцией И. А. Реддих. http://www.zorg-biogas.com

[6] А. П. Решетников, В. М. Меркелов and А. П. Решин. Теплогенератор многоцелевого назначения. Изв. Вузов. Лесной Журн, 2(320):С.138–140, 2011. Библиогр.: 5 назв.

[7] C. Mao, Y. Feng, X. Wang and G. Ren. Review on research achievements of biogas from anaerobic digestion. Renewable and Sustainable Energy Reviews, Elsevier, 45(C):540–555, 2015.

[8] I. Angelidaki, L. Treu, P. Tsapekos, G. Luo, S. Campanaro, H. Wenzel and P. G. Kougias. Biogas upgrading and utilization: Current status and perspectives. Biotechnology Advances, DOI: 10.1016/j.biotechadv.2018.01.011

[9] T. Al Seadi, B. Dominik, R. Rainer Janssen and B. Drosg. Biomass resources for biogas production. The Biogas Handbook Science, Production and Applications Woodhead Publishing Series in Energy, 19–51, 2013.

[10] IEA Bioenergy. Up-grading plant list. 2013. http://iea-biogas.net/plant-list.html.

[11] M. Persson and A. Wellinger. Biogas upgrading to vehicle fuel standards and grid introduction. Biogas Upgrading and Utilization, 2009.

[12] Q. Sun, H. Li, J. Yan, L. Liu, Z. Yu and X. Yu. Selection of appropriate biogas upgrading technology-a review of biogas cleaning, upgrading and utilisation. Renewable and Sustainable Energy Reviews, Elsevier, 51(C):521–532, 2015.

[13] A. Rafiee, K. R. Khalilpour, J. Prest and I. Skryabin. Biogas as an energy vector. Biomass and Bioenergy, 144:105935, 2021.

[14] Y. Mezmur and W. Bogale. Simulation and experimental analysis of biogas upgrading. AIMS Energy, 7(3):371–381, 2019. doi: 10.3934/energy.2019.3.371

[15] С. Я. Чернин and Ю. С. Парубец. Российский опыт внедрения биогазовых технологий для производства электрической и тепловой энергии. Источник: Журнал «новости Теплоснабжения», 08(132):2011 г., www.rosteplo.ru/nt/132 «Энергоэффективный город» № 7.3–1/40, www.energosovet.ru)

[16] П. Северилов. Биогаз для чайников. http://www.biogas.vn.ua http://www.agrobiogaz.ru

[17] А. М. Эфендиев. и др. Возможности энергообеспечения фермерских хозяйств на базе малых возобновляемых источников энергии. Теплоэнергетика, 2. С, 38–45, 2016.

[18] Н. И. Чернова. и др. Использование биомассы для производства жидкого топлива: современное состояние и инновации. Теплоэнергетика, 11. С, 28–35, 2010.

[19] З. Мамадалиева, Т. Каюмов, М. Султонов and Ф. Савриев. Сравнительный анализ применения биогазовых технологий и других источников энергии. © Портал ЭнергоСовет.ru – энергосбережение, энергоэффективность, энергосберегающие технологии 2006-2020
[20] W. R. Livingston. The Direct Co-Firing of Biomass at High Co-Firing Ratios, Doosan Babcock Energy, (2009), presented at Hamburg, June, www.ieabcc.nl/meetings/task32_Hamburg2009/cofiring/06%20Livingston.pdf. Moulton, B.W. (2009),
[21] Principles of Burner Design for Biomass Co-Firing. In: 34th International Technical Conference on Coal Utilization and Fuel Systems Clearwater, Florida, May 31, www.fwc.com/publications/tech_papers/files/TP_FIRSYS_09_02.pdf. ODE-Vl (2005).
[22] Biogas – Haalbare Kaart Voor Duurzame Landbouw (Biogas – a Viable Option for Sustainable Agriculture), ODE Vlaanderen, www.ode.be/images/stories/Brochures/bmg_fo_biogas_051201.pdf.
[23] S. Goldstein and X. Vitart. General comments about the efficiency of the iodine sulphur cycle coupled to a high temperature gas cooled reactor. ISJAEE, 3, 2004.
[24] Z. Struyk and G. Sposito. Geoderma, 102:329, 2001.
[25] B. H. Gu and J. Chen. Geochimica et Cosmochimica Acta, 67:3575, 2003.

A. Matyakubov

Chapter 4
Energy efficiency

Abstract: This chapter discusses the development of a highly efficient and inexpensive saltwater desalination installation using renewable energy sources such as solar energy.

The implementation of the installation proposed in this chapter is more rational and efficient than the previously existing installations. The results of studies comparing the operation of the existing installation and the proposed installation for desalination of water proved the distinguishing characteristics of the developed installation by its productivity. It was determined that the use of the proposed installation is possible in areas with the most favorable climatic conditions.

A detailed study using a theoretical approach has been carried out on solar water distillation technology. The study has revealed that this technology is very viable, capable of providing potable drinking water to rural communities remote from public potable water provision. The technology has demonstrated itself to be simple and cost-effective, making it affordable for the rural populace.

Keywords: Problems with drinking water, saltwater, desalination, the intensity of solar radiation in Turkmenistan, an autonomous distiller, water, desalination, solar panel, intensity of solar irradiation

4.1 Introduction

Protection of the environment, a careful approach to the nature, the preservation of its beauty and values for future generations are among the main tasks of the state. In this regard, a high rate of economic development and the creation of favorable conditions for Turkmenistan through the efficient and economical use of natural resources are envisaged.

Approximately 80% of the country's territory is occupied by the Karakum Desert. The main streams of rivers are located only in the southeastern regions of the country. The largest river, the Amu Darya, flows in the east of the country. Various problems arising from the lack of fresh water still remain as a serious issue in countries worldwide. Many scientific works in this direction explain the lack of fresh water and the search for solutions to this problem [1]. In addition to the availability of a rich supply of fresh water, high-quality drinking water is considered one of the necessary

A. Matyakubov, Bayramhan Street 62, Mary, Turkmenistan, e-mail: Amirhan31071989@mail.ru

https://doi.org/10.1515/9783110717655-004

conditions. Lack of clean drinking water can strongly affect human health. Many diseases in the world that affect human health are closely related to the consumption of low-quality fresh water. According to scientists' forecasts, it is assumed that in the near future, the probability of major problems in the world associated with the conquest of water resources will increase. In this regard, it is emphasized that the provision of fresh water remains a strategically necessary task for any state and should be at the center of attention for all countries [2, 3].

The problem of freshwater supply raises several questions. This issue becomes most critical when providing fresh water to densely populated areas, herding camps and military units. In this chapter, the solution to the problem of providing fresh water to places located far from the central power system using solar energy is considered. Solar energy is environmentally friendly, affordable, endless and a source of energy with a great future.

Use of renewable energy sources is widespread in all fields of the national economy and in water desalination as well. Moreover, it is important to note that using these installations operating on solar power at any populated areas remote from central heating and water supply systems is both profitable and eco-friendly.

It is considered that the solar cycle length in Turkmenistan is an average of 2,950–3,100 h, and the solar energy potential is up to 1.4 billion tons of conventional fuel [1].

The beneficial curve of water desalination in the Mary region during the year is equal to $\beta = 36°$. The seasonal curve in summer is $\beta = 23°$, in winter $\beta = 53°$ and in spring $\beta = 45°$. In the whole region of Turkmenistan, the activity of solar rays for 1 m^2 is 700–800 W [2].

4.2 Literature review

Desalination of water using solar energy has been a well-known method for a long time. Sailors in ancient Greece desalinated water in small containers by covering the containers with a transparent cover.

Ashkelon Desalination Plant [4] describes an efficient and economical battery-free photovoltaic system for seawater desalination using reverse osmosis. The system has a modest 2.4 kW photovoltaic battery and promises to deliver 3 m^3/day throughout the year in Eritrea, operating with seawater from a well (at 40,000 ppm).

Desalination using solar energy occurs with the help of a solar collector, where the heated liquid transfers its heat to the seawater in a heat exchanger, causing the seawater to evaporate in the condenser [5]. Desalination by solar energy is also discussed in [6], where various solar cells (photovoltaic), concentrating solar power, solar collectors and tower-type solar concentrators were used. A solar concentrator for desalination is the most efficient method, but it becomes necessary to track the sun's rays because the efficiency of the concentrator depends on the intensity of solar radiation.

Today, there are several ways to purify water, including direct osmosis, reverse osmosis and electrodialysis. The conservation of thermal energy using the heat-phase transition is described in [7]. It is also known that the production of fresh water requires a large amount of energy: desalination of 1,000 m^3 of fresh water per day requires 10,000 tons of oil per year [8]. Studying all the works about the desalination of solar energy, the authors made valuable remarks, where the intensity of solar radiation is high, i.e., in arid countries, there is a problem with drinking water. Water desalination technologies and their possible combination with solar energy were evaluated in [9]. The topic is of particular interest, especially for countries located in the southern Mediterranean belt, usually characterized by vast arid and isolated areas with little or no access to electricity from the national grid. A large number of systems that are used to convert seawater into fresh water suitable for human use is covered in [10]. It also covers many systems that can utilize renewable energy sources such as solar collectors, photovoltaic installations, solar ponds and geothermal energy. Both direct and indirect collection systems are included.

4.3 Structure and operating feature of saltwater desalination installations operating on the basis of solar energy

On the territory of Turkmenistan, each year solar energy amounts to 40×10^9 kJ or 1.4×10^9 tons of fuel equivalent, wind energy to 640×10^9 kW·h geothermal waters to 2.5 million tons of fuel equivalent. Additionally, there is a sufficient amount of biomass energy [11]. Solar power is the most cost-effective renewable energy source in the future. The climatic and geographical conditions of Turkmenistan make it possible to obtain solar energy and actively use it for generating electric and thermal energy on an industrial scale. The natural and climatic conditions of Turkmenistan are the most effective for the widespread use of renewable energy sources. The distribution of the energy potential of solar energy across the territory of Turkmenistan can be considered uniform due to its latitudinal location. The duration of daylight hours in the country in June is 16 h, in December 8–10 h, with 300 sunny days per year. The duration of sunshine is 2,500–3,100 h per year, and in summer – 320–400 h per month [12].

There are different kinds of saltwater desalination installations in the world. However, their operation causes problems connected with a lack of energy provision and leads to some definite issues hindering their implementation. Especially, remote areas located far from central energy provision systems face this problem [6–10].

One of the most popular methods of water desalination, i.e., the use of solar infrared rays, is the method of evaporation based on heat.

The main disadvantage of the aforementioned installation is that its efficient operation depends on daytime and the hot season, as shown in Figure 4.1.

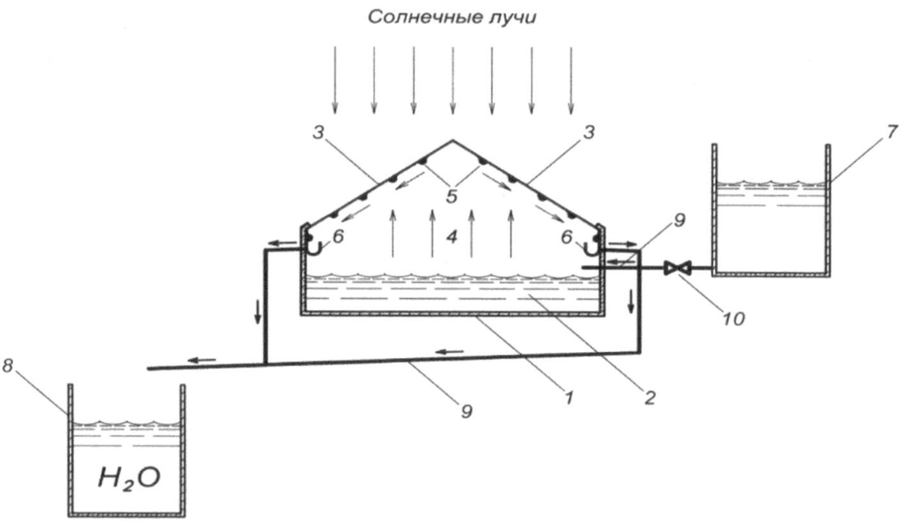

Figure 4.1: Typical solar water desalination installation with a capacity of 0.87 m^3 and its components: 1, external dimensions of water desalination installation; 2, reservoir of salty water (water pool); 3, glass covers installed at a 36° angle; 4, steam; 5, condensed drops; 6, channel for condensed drops accumulation; 7, spare tank for salty water; 8, tank for freshwater accumulation; 9, water pipe; 10, valve.

Various types of desalination installations have been tested under local conditions using solar energy. The performance of these desalination installations is related to efficiency, ambient temperature and working hours per day. Based on the results of research conducted on a conventional water desalination installation with a volume of 0.87 m^3, operating for 8 h a day at an ambient temperature of +35 °C, it was established at the Renewable Energy Sources Scientific and Production Center that this installation has the ability to produce 8–12 L of distilled water per day.

It is important to note that, by taking into account the achievements of innovative technologies and the use of the existing great possibilities of scientific advancements, it is possible to improve the methods of operation of previously known water desalination installations, as well as to increase their productivity. In this regard, the results obtained in this work can be used to effectively solve emerging problems in the provision of fresh water.

As can be seen from global practice, there are several types of installations for obtaining fresh water. Due to the fact that these installations are powered by electrical energy, there are some difficulties in putting them into operation. This problem occurs especially in settlements located far from the central distribution system.

When sterilizing water, a common method used is the application of thermal energy obtained from solar radiation. The main disadvantage of the previously known water desalination facility is that it produces distilled water only during the day and the warm season. At night and during cooler seasons, the facility loses its effectiveness [15].

This chapter presents research experiments conducted in a tightly closed water desalination installation. The installation consists of a desalination chamber (1) and a storage chamber (8). The volume of the desalination chamber is 0.14 m^3 and the volume of the freshwater storage chamber is 0.04 m^3. The surface of the installation is designed according to the optimal angle for effective collection of the intensity of solar radiation ($\beta = 36^0$) and is covered with a glass cover. The gap between the chambers is separated by a special divider (7). The distiller is also equipped with a water pump (4) for supplying salt water to the chamber. To increase the capacity of the distiller, a water heater (2) and a fan (5) are installed. They operate on a constant current. The fan (5) is used to circulate outside air through a pipe with a diameter of 16 mm inside the desalination tank, which in turn enhances the condensation process. The resulting fresh water is supplied through the valve (9) for use.

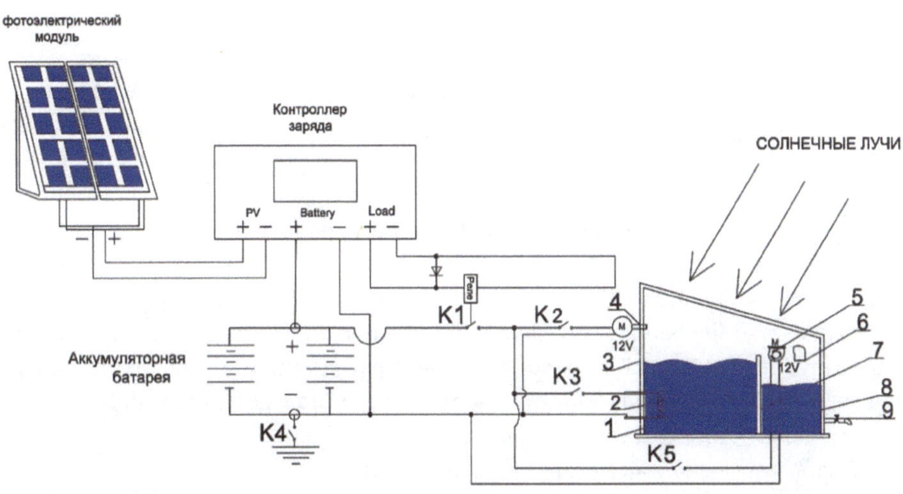

Figure 4.2: Connection diagram of the water desalination elements in the solar desalination installation: 1, desalinated water; 2, direct current water heater (12 V); 3, water desalination installation; 4, pump, operating at direct current (12 V), and part of a water desalination installation; 5, direct current (12 V) air exchange; 6, outlet part of the air supplied to the inside of the water desalination installation; 7, separator of the water desalination installation; 8, fresh water; 9, outgoing part of fresh water.

Water desalination installation operates 8 h in daytime, and the productivity of this installation depends on the outside air temperature. However, with the installation of an air exchanger, the circulation of outside air by pipes through installation leads to a change in the inside temperature to some extent and increases the chance of formation of condensed drops. During the day, to charge its accumulators with solar power, the solar panels of the water desalination complexes are turned to the south at a 36° angle [9]. The charge of the accumulators is adjusted by controllers. The water pump and air exchanger providing air circulation operating on the basis of direct current from these accumulators. At night, from 20:00 until 8:00 in the morning, a heater operating on direct current in the water desalination system heats the water, and fresh water flows continuously.

K1 is a general key used to control the installation entirely. The K2 key is an on-off switch for the water pump operating on direct current. The K3 key with 100 W power is an on-off switch for the water boiler. The K4 key is for the earth grounding of accumulators. The K5 key is an on-off switch for the air exchanger operating on direct current [16].

In this way, this installation operates continuously both day and night.

Several scientific research studies carried out in water desalination installations have determined that their efficiency depends on the outside temperature and the intensity of solar rays. As a result of these studies, it was determined that the efficiency of these installations can be increased by additionally installing photoelectric developers, heaters and air exchangers.

In the calculations, the desalination installation works 8 h during the day, taking this into account. During the day, the performance of this water desalination installation varies depending on the ambient temperature. Another feature of the distiller is the installation of a fan, which in turn leads to a change in the internal temperature in the pipeline (6) to a certain extent, which increases the occurrence of condensation.

The complex also includes solar panels, which are positioned at a horizontal angle of inclination of 36°. During the day, the solar panels of the water desalination complex charge the battery. The charge and discharge levels of the battery are controlled by the controller. The same battery also powers the water pump and fan. In the evening, from 8:00 pm until 8:00 am, the water heater heats the water, and thus the continuous process of water evaporation continues [17].

For the process of water evaporation, it is necessary:

- The temperature of mineralized water during the day varies from $t_1 = 18$ °C to $t_2 = 53$ °C.
- The volume of saline water, which is filled in the tank, is $\upsilon = 0.000236$ m^3.
- The specific heat capacity of water is equal to 4,191 J/kg.
- Average temperature is $t = \frac{t_1 + t_2}{2} = 35.5$ °C.

The amount of heat Q is equal to the product of the mass m of water and the temperature difference t:

$$Q = m \cdot t \tag{4.1}$$

We denote the initial water temperature as t_1 and the final temperature as t_2. Then

$$Q = m \cdot (t_2 - t_1) \tag{4.2}$$

Substitute numerical values:

$$Q = m(t_2 - t_1) = 14 \times (53 - 18) = 14 \times 35 = 490 \text{ J}$$

The productivity of any installation is determined by the time spent under optimal conditions. For a solar desalination installation, this time is the operating time when the Sun's rays are at right angles to the installation [13].

The intensity of solar radiation is calculated by the following expression:

$$E = H \cdot \cos i; \ \text{kW} \cdot \text{h/m}^2 \tag{4.3}$$

where H is the intensity of radiation incident on a horizontal surface ($\text{kW} \cdot \text{h/m}^2$) and i is the angle of inclination of the Sun's rays (beam) to this surface.

The cos i is calculated using the following formula:

$$\cos i = \cos \cdot (f - \beta) \cdot \cos \delta \cdot \cos \omega + \sin(f - \beta) \cdot \sin \delta \tag{4.4}$$

where δ is the angle of inclination; f is the width of the area (for the city of Mary, $f = 37°36$), β is the angle of inclination of the considered surface to the plane and ω is the hour angle of the Sun.

The angle of inclination is calculated using the following formula:

$$\boldsymbol{\delta} = 23,45 \cdot \mathbf{sin}\left(360 \cdot \frac{284 + \mathbf{n}}{365}\right) \tag{4.5}$$

where n is the ordinal number of the day of the year, counted from January 1.

As a result, the findings demonstrated the potential of using solar energy in the water desalination process. The results of scientific research indicate that utilizing the intensity of solar radiation is most effective in regions with many sunny days per year.

As a result of the conducted scientific research on water desalination using renewable energy sources, it was shown that the experimental data and mathematical models are not identical. This discrepancy affects the calculations of the productivity of installations, which will be used in different regions and under varying initial operating conditions.

As seen in connectivity, the efficiency of installation totally depends on the high level of outside temperature. On a typical sunny day over 24 h, a water desalination installation with a 0.15 m^3 capacity provides the opportunity to obtain 400 mL of fresh water.

Unlike outside temperature, the creation of high hot temperature and closed capacity in the evaporation chamber is a necessary condition to evaporate water simultaneously.

The dependence of the hot level of the outside temperature heat created in the evaporation chamber and the dependence of water desalination on time are shown in Figure 4.3.

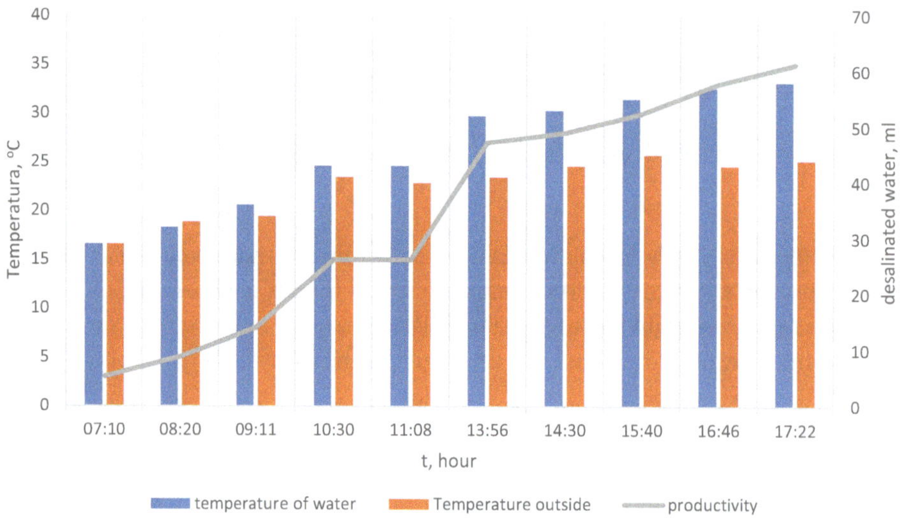

Figure 4.3: Dependence of the hot level of outside temperature heat created in the evaporation chamber and dependence of water desalination on time (November 9, 2020).

To improve the quality of scientific research work, i.e., to compare before and after evaporation conditions, examples were tested in the laboratory of the Sanitary and Epidemiology Service Center of the Mary region of the Ministry of Health and Medical Industry of Turkmenistan. The composition of the water before evaporation is given in Table 4.1.

Table 4.1: Composition of water before evaporation.

Title of the performance	Examination methods NTD	Cost of performance according to the criteria	Determined actual cost	Performance criteria
1 Water odor 20 °C and when heated at 20 °C, not much in points	TSS-837-2016	2	1/1	Match
2 Color, not high in degree	TSS-3351-74	20	5	Match
3 Muddiness, by standard measurement mg/dm	TSS-3351-74	1.5	319	Not match
4 Hydrogen indicator, pH	TSS-837-2016	6–9	7.8	Match

Table 4.1 (continued)

	Title of the performance	Examination methods NTD	Cost of performance according to the criteria	Determined actual cost	Performance criteria
5	Hydrocarbonate oxidation (HCO_3), mg/dm	TSS-23268.3–78	–	5.2	–
6	General sugary, mg-ekw/dm^3, little	TSS-4151-72	7.0	21.5	Not match
7	Ions of calcium (Ca^{2+}), mg/dm^3	TSS-3354-74	–	286	–
8	Ions of magnesium (Mg^{2+}), mg/dm^3	TSS-3354-74	–	87.5	–
9	Chlorides, mg/dm^3, a little bit	TSS-955-2012	350	340	Match
10	Sulfates, mg/dm^3, a little bit	TSS-4389-72	500	135	Match

As given in the schedule, it is evident that the composition of the water does not meet some requirements. For instance, the muddiness of the water is 217 times higher than the allowed norms, and the sugar content is 3 times higher than the allowed norms. Additionally, the amount of calcium, magnesium ions, and chloride in the salty water is very high. Therefore, such water composition is unsuitable for drinking, as certified by the results obtained from the laboratory.

Water composition of evaporated water by solar installation is given in Table 4.2.

As seen in the schedule, the water composition of evaporated water from the water desalination installation over 24 h becomes much cleaner. Specifically, the muddiness

Table 4.2: Composition of water after evaporation.

	Title of the performance	Examination methods NTD	Cost of performance according to the criteria	Determined actual cost	Performance criteria
1	Water odor 20 °C and when heated at 20 °C, not much in points	TSS-837-2016	2	1/1	Match
2	Color, not high in degree	TSS-3351-74	20	5	Match
3	Muddiness, by standard measurement mg/dm	TSS-3351-74	1.5	1.16	Match
4	Hydrogen indicator, pH	TSS-837-2016	6–9	7.8	Match

Table 4.2 (continued)

Title of the performance		Examination methods NTD	Cost of performance according to the criteria	Determined actual cost	Performance criteria
5	Hydrocarbonate oxidation (HCO_3), mg/dm	TSS-23268.3–78	–	1.1	–
6	General sugary, mg-ekw/dm^3, little	TSS-4151-72	7.0	0.4	Match
7	Ions of calcium (Ca^{2+}), mg/dm^3	TSS-3354-74	–	4.0	–
8	Ions of magnesium (Mg^{2+}), mg/dm^3	TSS-3354-74	–	24.3	–
9	Chlorides, mg/dm^3, köp däl	TSS-955-2012	350	60	Match
10	Sulfates, mg/dm^3, köp däl	TSS-4389-72	500	90	Match

of the water is reduced by 275 times, hydrocarbonate oxidation (HCO_3) by 4.7 times, general impurities by 54 times, calcium ions (Ca^{2+}) by 71.5 times, magnesium ions (Mg^{2+}) by 3.6 times, chloride content by 5.6 times and sulfate content by 1.5 times. Therefore, water with such a composition is safe to drink, as certified by the results obtained from the laboratory.

4.4 Discussion

As a result of the experiments, it can be concluded that the use of solar energy makes it possible to develop a high-performance water desalination installation at a low cost, which can be effectively used in the conditions of Central Asia.

4.5 Conclusion

As a result of scientific research carried out in water clarification by means of renewable energy sources such as solar power, this installation offers water desalination opportunities in various weather conditions:

1. Based on the solar water desalination installation developed at the research and production center and in comparison with previously existing types, the advantages of the recommended installation have been proven. High reliability and the possibility of using this technology in warm climate zones have also been demonstrated.

2. The water desalination installation, the structure of which is made of a glass cube, has shown high desalination performance. During the experimental work, high reliability was demonstrated as a result of assembling the structure from the most transparent and fire-resistant glass.
3. The possibility of using a desalination installation during the day, in both sunny and cold climates with the use of an additional water heater, has been proven.
4. In scientific work, an important scientific and technical problem in the field of water supply has been solved, i.e., water desalination installation has been developed, which can be used in desert areas far from settlements.

Consequently, utilizing solar power in the water desalination process yielded significant results, and the offers are as follows:
1. This installation can produce 400 mL of clarified drinking water during a sunny day (24 h).
2. On a sunny day, 120 mL of clarified drinking water can be produced from 07:00 in the morning until 17:00.
3. During the night, 180 mL of clarified and desalinated drinking water can be produced from 18:00 to 06:00 in the morning.
4. If water is aimed directly for home use without a photoelectric converter installation, then its nominal power is 100 W. If electric heaters operating on 220 V and other electric devices are installed, then 1.5 kW per hour will be consumed for 11 h, which is equal to 0.05 manat.
5. The cost of this installation is 1,500 manat, and its pay-off period is 1.5 years (the cost of 400 mL of drinking water is considered to be 3 manat in shops).

References

[1] A. Jumayev and H. Soltanow. Basement of Energy saving.-A.. Science, 2018.

[2] S. Nazarow and A. Jumaýew. Opportunities of solar power use in regions of Turkmenistan. Science and Technology in Turkmenistan, 2019, 6.

[3] IDA Desalination Yearbook 2006–2007. Water Desalination Report, Global Water Intelligence and International Desalination Association.

[4] Ashkelon Desalination Plant, Seawater Reverse Osmosis (SWRO) Plant, Water technology net. http://www.water-technology.net/projects/Israel (21 November 2011) Renewable desalination market analysis in Oceania, South Africa, Middle East & North Africa, 2010. www.aquamarinepower.com (November 2011).

[5] H. Mahmoudi, O. Abdellah and N. Ghaffour. Capacity building strategies and policy for desalination using renewable energies in Algeria. Renew Sustain Energy Rev, 2008, in press.

[6] US Energy information administration International Energy Outlook 2010 – highlights. http://www.eia.doe.gov (26 November 2011).

[7] M. T. Ali, H. E. S. Fath and P. R. Armstrong. A comprehensive techno-economical review of indirect solar desalination. Renewable and Sustainable Energy Reviews, 15:4187–4199, 2011.

[8] I. C. Karagiannis and P. G. Soldatos. Water desalination cost literature: Review and assessment. Desalination, 223:448–456, 2008.
[9] E. Tzen. Renewable energy sources for desalination. In: Paper presented at Workshop on Desalination Units Powered by RES. Athens, 2006.
[10] K. Quteishat and M. Abu Arabi. Promotion of Solar Desalination in the MENA Region Report. Middle East Desalination Research Centre, 2004.
[11] M. Thomson and D. Infield. A photovoltaic-powered seawater reverse-osmosis system without batteries. Desalination, 153:1–8, 2002.
[12] K. Al Subaie. Precise way to select desalination technologies Desalination 206:29–35, 2007. Elsevier.
[13] M. I. Ahmed, M. Hrairi and A. F. Ismail. On the characteristics of multistage evacuated solar distillation, Renewable Energy, 34:1471–1478, 2009. B. A. Jubran, M. I. Ahmed, A. F. Ismail and Y. A. Abakar. Numerical modeling of a multi stage solar still, Energy Convers. Manage, 41:1107–1121,2000.
[14] Pascale Compaina* Solar Energy for Water desalination. In: 1st International Symposium on Innovation and Technology in the Phosphate Industry [SYMPHOS 2011].
[15] Kalogirou. Seawater desalination using renewable energy sources. Progr Energy Combustion Science, 31:242–281, 2005.
[16] G. Fiorenza and V. K. Sharma. Techno-economic evaluation of a solar powered water desalination plant. Energy Conversion and Management, 44:2217–2240, 2003.
[17] Kalogirou. Seawater desalination using renewable energy sources. Progr Energy Combustion Science, 31:242–281, 2005.

S. Nazarov

Chapter 5 Latent renewable energy in Turkmenistan

Abstract: Among the main tasks facing the international community today is the expansion of the use of renewable energy sources, which are associated with climate change and negative environmental impacts.

Turkmenistan has significant potential for renewable energy sources, the development of which will bring positive environmental, economic, and social benefits. Given this potential, Turkmenistan has adopted a number of state programs and strategies for the development of renewable energy sources.

The peculiarities of the climatic and geographical conditions of Turkmenistan open up huge opportunities for the development of renewable energy in the country. These opportunities will help improve energy supply to consumers far from the central grid and reduce carbon dioxide emissions into the environment. Of particular interest are solar and wind energy. The potential of solar energy per year is equivalent to 1.4 tons of fuel, and the potential of wind energy is 640 billion kWh.

According to the research results, it can be said that there are about 2,950–3,100 h of sunshine in Turkmenistan, which provides significant opportunities for their use in the national economy for the production of electricity and heat. Additionally, this scientific work notes the available potentials of geothermal waters and hydroenergy are noted

Keywords: Solar energy, solar radiation intensity in Turkmenistan, optimal angle of inclination of the solar PV, hydroenergy, geothermal power

5.1 Introduction

Sustainable energy and energy-efficient technologies have been attracting much attention in recent years due to the soaring energy crisis and environmental problems associated with the depletion of natural resources. Electricity generation from the burning of fossil fuels is a major source of greenhouse gas emissions, leading to global warming. Sustainable energy resources such as solar, wind, biomass, hydrogen, geothermal, ocean and hydropower not only help conserve fossil resources for future generations but are also considered clean sources of energy that are constantly replenished.

S. Nazarov, Bayramhan Street 62, Mary, Turkmenistan, e-mail: marytdei@mail.ru

https://doi.org/10.1515/9783110717655-005

Development of renewable energy sources (RESs) in Turkmenistan is primarily important for electricity generation, heating and cooling provision.

Peculiarities of the fuel energy complex of Turkmenistan include the presence of large natural gas reserves. Moreover, there are many petroleum reserves in Turkmenistan. The quantity of fossil hydrocarbons increases in comparison to the demand of internal consumers, and it has great export potential.

Local electric stations burn natural gas, fuel oil and diesel fuel to generate electricity. All types of hydrocarbon fuel mining and fossil mining are local productions. Moreover, Turkmenistan has great potential for RESs such as solar and wind.

5.2 Literature review

A general description of the use of RESs by sectors of the US economy is given in [1], where 2% of the total consumption came from geothermal sources, 11% from photovoltaic (PV) solar stations, 26% from wind power plants and 4% from biomass energy, in 2021. Considering that the United States is increasing its energy potential every year, RESs will help increase exports and reduce carbon dioxide emissions.

A report in [2] states that for the United Arab Emirates, RESs are cost-effective. Ramping up renewables to 10% of the country's total energy mix and 25% of total power generation could generate annual savings of USD 1.9 billion by 2030 through the avoidance of fossil fuel consumption and lower energy costs. With health and environmental benefits factored in, the transition to renewables could generate additional net annual savings of USD 1 billion to as much as USD 3.7 billion by 2030.

This report is a joint effort by the International Renewable Energy Agency (IRENA), the UAE-based Masdar Institute, and the UAE Ministry of Foreign Affairs' Directorate of Energy and Climate Change. The analysis of the UAE's renewable energy prospects is part of REmap 2030, IRENA's roadmap for doubling the share of renewables in the global energy mix.

An analysis was carried out in [3] that in Russia in 2010, the most demanded type of renewable source of heat energy used in the sectors of industry and housing and communal services (including district heating) was bioenergy, while hydropower dominated in electricity production. In 2010, hydropower and bioenergy accounted for 70% and 30% of the total final energy consumption (0.6 exajoules, EJ), respectively, of renewable energy. In the same year, the share of renewable energy in the total final energy consumption in Russia was 3.6%. By the end of 2015, the total installed electric capacity of facilities operating on the basis of the use of RES reached 53.5 GW, which was about 20% of the total installed electric capacity in Russia (253 GW). Hydropower accounted for almost all of the installed capacity – 51.5 GW, followed by bioenergy at 1.35 GW. The installed capacities of solar and wind power plants amounted to 460 MW and 111 MW, respectively. Based on the results of consultation with the Russian government and relevant stakeholders, the report identifies four main driving forces that, according to Russia, will accelerate the introduction of renewable energy into the structure of the Russian energy sector: economic activity and job creation, development of science and technology, energy supply to isolated energy regions and improving the quality of the environment. In 2013, it was reported that renewables made up only 0.2% of Iran's installed generation capacity. The total installed capacity of PV power plants was under 90 kWh, most of which supplied electricity to street lighting. At present, renewables contribute little more than 1% of Iran's total primary energy consumption, with the large majority of this made up of long-standing hydroelectric power. Iran is a large country with diverse climate and topography. In terms of wind resources, estimates of the total potential capacity in Iran range from 30 GW to 100 GW. Solar resources are also abundant, with much of the country having in excess of 300 days of sunshine annually. It is thought that the country's geology may support the deployment of significant amounts of geothermal energy. Finally, there is potential for using a substantial number of waste streams for energy generation purposes [4].

Pillai and Banerjee [5] examined the state and potential of various RESs (except biomass) in modern India. The growth trend of renewable energy in India is noted, and a diffusion model is established as a basis for setting targets. The diffusion model follows past trends in wind, small hydropower and solar water heating and is used to set future targets. Economic viability and reductions in greenhouse gas emissions are assessed for each option. Some RESs have high growth rates, such as the production of wind turbines, PV modules and solar water heaters (concentrated solar power). New technologies such as tidal, Ocean Thermal Energy Conversion (OTEC), solar thermal power plants and geothermal power plants are under study for use at the state level. Barrington and Ouliaris [6] gave an assessment of Canada's RESs. Their assessments included wind, solar, hydro, tidal, wave and geothermal energy, with limited discussion of bioenergy. For comparison, we have also broken down the current energy demand in each province according to the categories that are relevant to households. The authors found that overall, with current technology, Canada can more than just meet its energy needs using RESs, two-thirds of which will come from onshore and offshore winds, with much of the remainder coming from hydroenergy. However, they found large differences between provinces in the mix and scope of renewable potential.

Yusaf et al. [7] examine the potential of renewable resources in Australia today, as well as the magnitude of their current and future energy consumption. Based on the results of this scientific work, specific proposals and recommendations were given on the use of solar, wind and wave energy in the ocean.

Enjavi-Arsanjani et al. [8] used Global Horizontal Irradiance (GHI), measurement data for 21 cities in Iran and selected 6 urban areas with higher solar potential. Modeling was carried out for a 100-MW parabolic trough power plant with 6 h of heat accumulation using software. The research results showed that the site in the Shiraz area has the highest potential to generate electricity, while Bandar-e Abbas (at the Persian Gulf) has the lowest potential:

- From the sites analyzed, Shiraz, Yazd and Kerman have areas with higher solar radiation, as illustrated in the solar GIS map. All these areas suffer from water shortages, and therefore dry cooling towers are the best condensing systems for these locations.
- The vast arid land around these areas, as shown in the solar GIS map, is the main advantage of the above sites.
- Fossil fuel consumption can be reduced considerably in these areas, and CO_2 emissions can be decreased in line with sustainable development programs.
- Both Iran and Algeria are in the sun-belt region of the world, and computations indicate that there is a good opportunity to harness solar energy for electricity generation in both countries.

Karataye et al. [9] also provide the total energy consumption, which was produced using RESs, whose share in Kazakhstan is less than 1%. Additionally, Kazakhstan has a national strategy in the field of low-carbon energy, aimed at increasing the share of RESs in electricity production to 50% by 2050.

5.3 Materials and methods

5.3.1 Solar energy

Due to its climate conditions, the region of Turkmenistan is considered the most appropriate area for solar energy generation. The annual duration of sunlight is 2,500–3,100 h. In the hottest month of the year, July, the amount of solar energy falling on a 1 m^2 horizontal surface per day is 6.4–7.5 kW. Generally, the solar energy potential of Turkmenistan can be equivalent to 4×10^{15} kJ or 1.4 milliard tons of conventional fuel per year [10].

Calculations of monthly solar radiation falling on a surface turned in one definite directed corner, according to the horizontal plane, can be performed using the Liu-Jordan method [11]:

$$E_a = R \cdot E_i \tag{5.1}$$

where E_a is the average monthly cost of total radiation falling on a specific directed corner during the daytime, E_i is the average monthly cost of total radiation falling on a horizontal surface during the daytime, R is the balance of the average monthly cost of total radiation falling on a specific directed corner and a horizontal surface during the daytime.

In the experimental scientific research field of the Scientific and Production Center "Renewable Energy Sources" at the State Energy Institute of Turkmenistan, the activity of Sun's rays was measured continuously using special devices. Measurements of the radiation quantity falling on the water desalination unit carried out horizontally and at 36° angle are shown in Figure 5.1. The date of measurement was July 8, 2020, and the coordinates of Mary city in north latitude is 37.6° and east latitude is 61.8°.

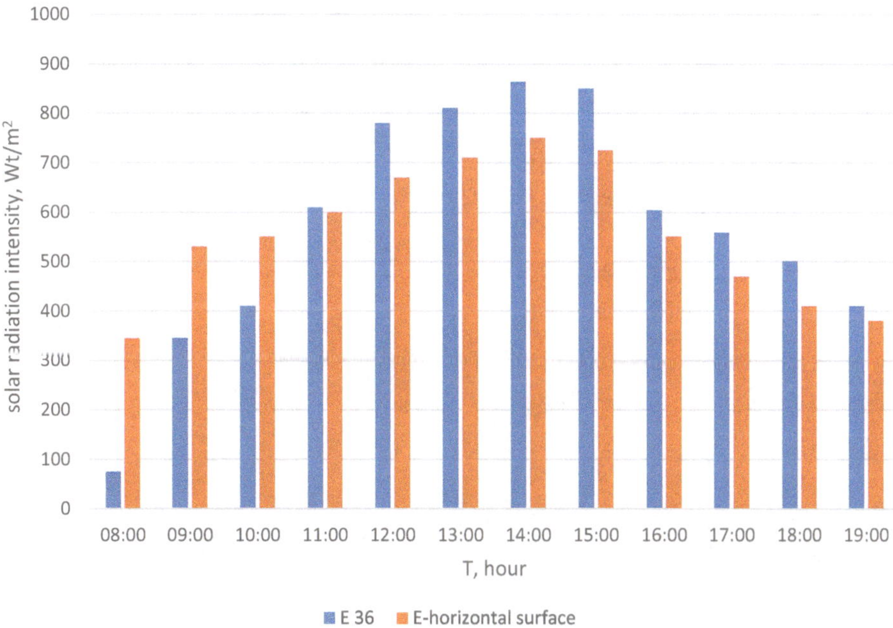

Figure 5.1: Change in active Sun's rays falling on the water desalination unit horizontally and at 36° angle by hours during the day (July 8, 2020).

According to the measurements taken on July 8, 2020, the amount of energy falling on the horizontal surface is large in the first part of the day. However, at 11:00, in the second part of the day, Sun's rays fall more intensely on the consumer, curved at 36° angle. The results from October 26, 2020, show that as the Sun comes closer to the Earth, the intensity of Sun's rays is higher compared to the results from July 8, 2020. Therefore, installing a water desalination unit at 36° curve throughout the year in the conditions of the Mary region is more effective.

An average monthly (July 7, 2020, to August 6, 2020) amount of the intensity of Sun's ray is shown in Figure 5.3.

As shown in Table 5.2 the balance, the amount of energy falling during a month is 560–590W/m^2, and it is evident that the solar energy potential in the territory of Mary city is great.

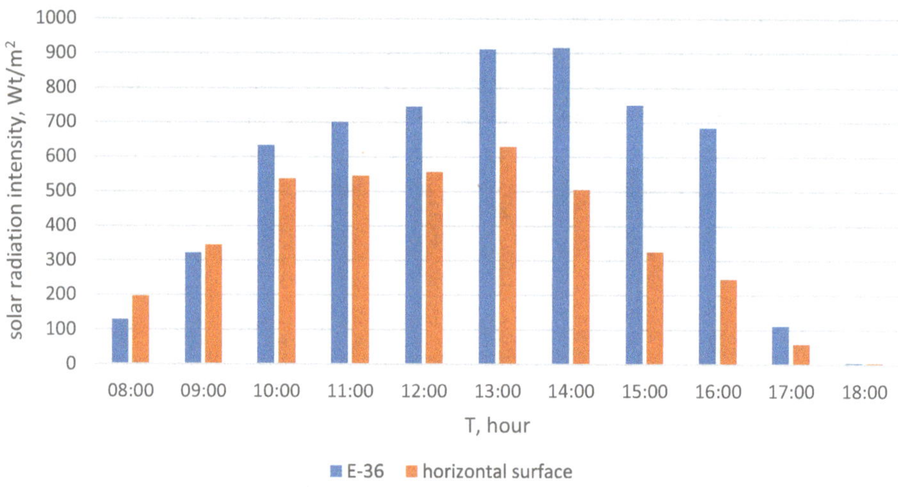

Figure 5.2: Change in active Sun's rays falling on the water desalination unit horizontally and at 36° angle by hours during the day (October 26, 2020).

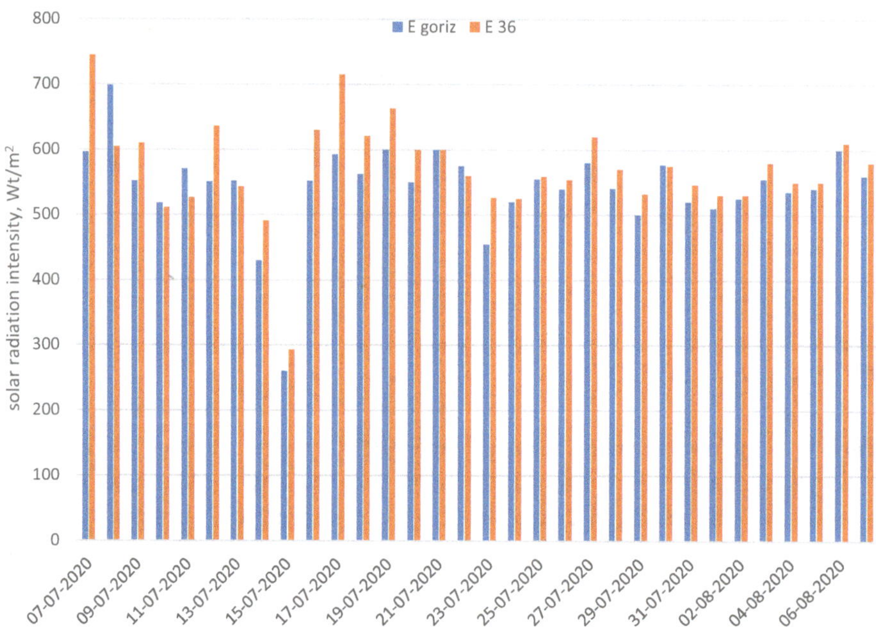

Figure 5.3: Monthly fall amount of Sun's rays in Mary city during a month.

5.3.2 Hydroenergy

There are a few low-water rivers due to the dry climate of the country, the small amount of precipitation and the Turkmen mountains (Kopet Dagh) with no ice.

Amu-Darya is the only significant river entering from the east of the country, fed by Pamir ice. Currently, there is no river network in the main territory of the country. Of the 179 registered water bodies, only 40 are rivers with stable water, 12 are seasonal rivers that dry up and more than 100 are heavy rain channels where stormy rainfalls last from a few hours up to 3 days and nights.

One of the ways to develop RESs in Turkmenistan is by using the energy of small streams through the installation of mini- and micro-hydroelectric stations in mountainous areas. In the water-scarce conditions of Turkmenistan, great attention should be paid to small rivers, which are the most vulnerable parts of river systems. Therefore, considering the opportunities for hydroenergy power development of small rivers, the risks and benefits of installing hydropower stations on these rivers should be thoroughly evaluated. The development of mini-hydropower should be connected with the improvement of small river conditions. Constructed dams and water reservoirs should not only ensure the presence of small rivers but also contribute to their development. Taking into account the water demand of neighboring agricultural and public enterprises, water protection projects should be developed.

Scientific experimental work on the opportunities for mini-hydropower development should be continued along with the protection and development of small rivers. At the same time, technical and economic feasibility studies and cost recovery should be taken into account.

The first hydroelectric power station in the region of Turkmenistan was the Hindukush Power Station, built on the banks of the Murgab River in 1913. The construction of this hydroelectric power plant was entrusted to the company "Tami and Deichman," which has a complex of construction machines, equipment and experienced specialists. The equipment of the Austro-Hungarian company "Hans" was installed at the station.

The Hindukush HPP has three hydraulic turbines with a capacity of 400 kW each, and a total capacity of 1.2 MW.

The shaft of the hydroturbines at the Hindukush hydropower plant rotates under a powerful stream of water coming from a dam built on the Murgab River.

The electricity generated by the hydroelectric station is transmitted through 38.7 km of power lines to consumers such as a creamery, a cotton processing plant, a soap factory and other facilities in the city of Bayramaly.

The distance between the power transmission supports (towers) ranges from 25 to 43 m, and their height, depending on the place of installation, varies from 7 to 14 m. These overhead power lines were constructed manually using various supports and cables.

There are many items and pieces of equipment at the power plant that were installed 100 years ago. These include three hydraulic turbines, signal lamps that still work flawlessly, a set of keys of different sizes supplied by the contractor when the station was commissioned and street lights [12].

5.3.3 Geothermal power

The main advantages of geothermal energy are stability and absolute independence from environmental conditions, time and seasons.

In Turkmenistan,, the general heat energy generation of studied thermal waters per year is equal to 17.5 million Gcal or 2.5 tons of conventional fuel. Its extraction amount (debit) is 1.3 million m^3 per day [13].

The temperature of geothermal water from the well is about 35–50 °C. A temperature of 150 °C or higher is appropriate for generating electric energy. A temperature of 50 °C is required for heating and hot water provision. The temperature of the soil increases very slowly with depth; as a rule, the geothermal gradient for 1 km is just equal to 30 °C.

5.4 Recommendations for the renewable energy development in Turkmenistan

In the energy system of Turkmenistan, gas-turbine power stations are dominant. This provides opportunities for the systematic introduction of electric energy generators operating on the basis of combined RESs to generate electricity. From a technical point of view, there is no problem in developing renewable electric energy in the country.

Conditions for developing the use of RESs are profitable from an economic point of view. The cost of natural gas used to generate electricity is very low. As a result, the cost of electricity generated in local power stations is low as well. The cost of electric energy generated using RESs is becoming and is forecasted to be much cheaper in the future. Nowadays, in public sales, investors started to offer equal or higher prices for constructing stations operating on the basis of RESs compared to those operating on the basis of fossil fuels.

Development of the three main profitability aspects of using RESs in electricity generation is as follows:

1. Export
2. Local consumption in the territories of the combined cycle energy system
3. Electricity provision for long-distance consumers

Nowadays, Turkmenistan is an exporter of various energy resources. From a state perspective, exporting goods and services with higher added value than raw materials is beneficial. Producing goods and services within the country creates new jobs, improves material and technical conditions, enhances the living standards of the population and advances manufacturing to new technical levels. When the costs of transportation networks for hydrogen, fossil fuels and electric energy are reasonable, exporting electric energy becomes profitable.

Energy export capacity of Turkmenistan to neighboring countries with low energy potential is high. Considering the comparatively low expenses to balance and guarantee the cost of solar and wind energy generation, as well as electric stations operating on combined cycle RESs, developing RESs can be profitable for Turkmenistan to export electric energy in the long term. Conducting scientific researches and applying a scientific approach to solving this problem are rational steps in defining the export profitability of electricity generation based on RESs.

In the domestic market, the cost of electricity generated by local gas in gas turbine power plants is lower than that of electricity generated by solar and wind energy. However, soon investors may receive competitive price offers for electricity generation based on RESs.

According to the results of scientific research conducted by scientists at the State Energy Institute of Turkmenistan, using RESs to generate electricity for consumers (frontier post stations, shepherd villages, farmers, etc.) located far from electric networks is the most profitable solution to the energy provision problem. Constructing small energy networks to provide electricity to consumers located far distances requires significant investments and operating expenses. Moreover, in this case, energy transformation and distribution over such long distances lead to energy loss, which is unprofitable. Providing electricity to remote locations through local power plants operating on RESs, along with mini- and micro-networks for their distribution, is the most modern and profitable method and technological solution to this problem.

5.5 Conclusion

In the field of heat provision and domestic hot water systems, the use of solar power is profitable for heating water and industrial stations. Heat absorbers have great potential for providing both heat and cold. Additionally, there is an opportunity to prepare food cleanly in remote places using RESs.

In the far future, the use of energy generated by RESs may be applied in the transportation field, in powering electric means of transport, and biogas may be used as a motor oil.

List of activities in support of RESs development is given in the first place.

Main obstacle in solar energy development is dusty and stormy weather of Turkmenistan which can reduce energy generation. This is especially true in the spring when dust and haze from sand particles in the air can lead to reduction in energy generation.

Solar energy generation can be high when installed in desert areas where there are no territorial problems, and central electricity provision is available.

References

[1] U.S. Energy Information Administration, Monthly Energy Review, Table 1.3.

[2] https://www.irena.org/publications/2015/Apr/Renewable-Energy-Prospects-United-Arab-Emirates

[3] Remap 2030 Renewable Energy Prospects For The Russian Federation April 2017 Working Paper, ISBN 978–92-9260-022–8 (PDF) https://www.irena.org/media/Files/IRENA/Agency/Publication/2017/Apr/IRENA_REmap_Russia_paper_2017.pdf

[4] Renewable energy projects in Iran: solar focus, file:///C:/Users/user/Downloads/Iran%20Solar%20Brochure.pdf

[5] I. R. Pillai and R. Banerjee. Renewable energy in India: Status and potential. Energy, 34(8):970–980, 2009.

[6] C. Barrington and L. M. Ouliaris. The renewable energy landscape in Canada: A spatial analysis. Renewable and Sustainable Energy Reviews, 75:809–819, 2017.

[7] T. Yusaf, S. Goh and J. A. Borserio. Potential of renewable energy alternatives in Australia. Renewable and Sustainable Energy Reviews, 15(5):2214–2221, 2011.

[8] M. Enjavi-Arsanjani, K. Hirbodi and M. Yaghoubia. Solar energy potential and performance assessment of CSP plants in different areas of Iran. International Conference on Concentrating Solar Power and Chemical Energy Systems, Solar PACES 2014, Energy Procedia, 69:2039–2048, 2015.

[9] M. Karatayev, S. Hall, Y. Kalyuzhnova and M. L. Clarke. Renewable energy technology uptake in Kazakhstan: Policy drivers and barriers in a transitional economy. Renewable and Sustainable Energy Reviews, 66:120–136, 2016.

[10] Confirmed by decree of the President of Turkmenistan «national Strategy of Renewable Energy Sources in Turkmenistan by 2030 y». Ashgabat, 2020ý

[11] A. Jumayew. Scientific-technical and methodical analysis of development of solar power sources in Turkmenistan. Educational Book for Students of HEI.-A., 2016.

[12] Official site of Ministry of Energy Turkmenistan, http://www.minenergo.gov.tm/ru/node/13

[13] A. Matyakubov and A. Muhammetniyazov. Potential of geothermal energy in Turkmenistan. In: Materials of the XVII International scientific and practical conference. Minsk, 394, March 12, 2021.

Saliha Demirbüken and Vilda Purutçuoğlu*

Chapter 6
Approximate stochastic simulation algorithms

Abstract: Stochastic Simulation Algorithms (SSAs) utilize Monte Carlo methods to precisely model molecular population dynamics in alignment with the chemical master equation (CME). Despite their accuracy, SSAs can be computationally intensive, often requiring extensive simulation of reaction events and thousands of sample paths to accurately characterize probability distributions. These methods are particularly suited for systems with finite molecular populations. To address the computational challenges, approximate SSAs offer a practical alternative, compromising exactness for efficiency. These approximations are grounded in Markov processes and the leap condition, which assumes minimal change in propensity functions over discrete time intervals. The approximate SSAs include the Poisson τ-leap, Langevin τ-leap, estimated midpoint techniques, binomial τ-leap, modified Poisson τ-leap method, and newer approaches such as the τ-selection procedure and approximate Gillespie algorithm. Additionally, advances in statistical analysis, using second and third order truncated Taylor series expansions, enhance the understanding of distribution characteristics by providing insights into variance and covariance, leading to more precise confidence intervals compared to previous studies. This chapter explores the theoretical foundations, practical applications, and computational trade-offs of both exact and approximate SSAs, offering a comprehensive overview for researchers in stochastic process.

Keywords: Approximate stochastic simulation algorithms, leap condition, biological networks

6.1 Introduction

The stochastic simulation algorithm (SSA) can be defined as an application of Monte Carlo techniques to evaluate numerically the time trajectories of molecular populations in exact accordance with the chemical master equation (CME) [1, 2]. Although these methods provide exact results, they require a lot of time to produce outcomes because

Acknowledgment: The authors thank the METU research grant (BAP project no. 10282) for the support.

***Corresponding author: Vilda Purutçuoğlu,** Department of Statistics, Middle East Technical University, 06800 Ankara, Turkey, e-mail: vpurutcu@metu.edu.tr
Saliha Demirbüken, Department of Mathematics, Gebze Technical University, 41400 Kocaeli, Turkey, e-mail: sdemirbuken@gtu.edu.tr

https://doi.org/10.1515/9783110717655-006

they necessitate to simulate every reaction event [3]. Consequently, algorithms based on Monte Carlo simulations need thousands, tens of thousands or even hundreds of thousands of sample paths to acquire an accurate picture of the underlying probability distributions [4]. Furthermore, these methods are preferable for distinctly finite chemical systems [5]. For more information, we recommend the following references [1, 5–8].

When considering these issues, approximate SSAs can be chosen efficiently despite the loss of exactness. These methods depend on the leap condition, and the mathematical details of which are explained in the following section.

6.1.1 Leap condition

Consider a general form of the chemical reaction as follows:

$$m_1A_1 + m_2A_2 + \cdots + m_rA_r -> n_1B_1 + n_2B_2 + \cdots + n_pB_p \tag{6.1}$$

where r is the number of reactants and p is the number of products. A_i refers to the ith reactant molecule and B_j refers to the jth product molecule. Additionally, m_i means the number of molecules of A_i consumed in a single reaction step, while the subindex i is $i = 1, 2, \ldots, r$, and n_j represents the number of B_j produced in a single reaction step for $j = 1, 2, \ldots, p$. Furthermore, the coefficients m_i and n_j are called stoichiometries [8].

Basically, consider the following system [5]:

$$\begin{aligned} &R_1 : Y_1 + Y_2 \overset{k_1}{\rightarrow} Y_3 + Y_4 \\ &R_2 : 2Y_1 \overset{k_2}{\rightarrow} Y_1 + Y_2 \\ &R_3 : Y_1 \overset{k_3}{\rightarrow} Y_2 \end{aligned} \tag{6.2}$$

where R_j ($j = 1, 2, \ldots, p$) demonstrates the reaction channel. In this statement, the quantity of each chemical is considered as a concentration. Lastly, $[Y]$ refers to the notation for the concentration of a chemical species Y, as shown in eq. (6.3) [8].

The ordinary differential equations (ODEs) for system (1.2) can be represented by

$$\begin{aligned} \frac{d[Y_1]}{dt} &= -c_1[Y_1][Y_2] - c_2[Y_1] - c_3[Y_1] \\ \frac{d[Y_2]}{dt} &= -c_1[Y_1][Y_2] + c_2[Y_1] + c_3[Y_1] \\ \frac{d[Y_3]}{dt} &= c_1[Y_1][Y_2] \\ \frac{d[Y_4]}{dt} &= c_1[Y_1][Y_2] \end{aligned} \tag{6.3}$$

where c_1, c_2 and c_3 are the rate constants that show the speed of each reaction (j = 1, 2, 3). Then, v_1= (−1, −1, 1, 1), v_2 = (−1, 1, 0, 0), v_3= (−1, 1, 0, 0).

Here, the vector of v_{μ_i} indicates the change in the S_i molecular population caused by the occurrence of one R_μ reaction [5]. Moreover, the system can be shown in matrix form as follows:

$$\frac{d}{dt}\begin{bmatrix}[Y_1]\\ [Y_2]\\ [Y_3]\\ [Y_4]\end{bmatrix} = \begin{bmatrix}-1 & -1 & -1\\ -1 & 1 & 0\\ 1 & 0 & 0\\ 1 & 0 & 0\end{bmatrix}_{4\times3} \begin{bmatrix}c_1[Y_1][Y_2]\\ c_2[Y_1]\\ c_3[Y_2]\end{bmatrix}_{3\times1} \tag{6.4}$$

where the (4 × 3)-dimensional matrix refers to the stoichiometry matrix, V, of the reaction system [8]. The set of equations (6.3), also known as the reaction rate equations, is a set of coupled, first-order, ODEs traditionally used to define the time evolution of a well-stirred chemical system [3].

Besides being the reaction rate constants k_i in deterministic systems, the stochastic system has stochastic rate constants c_i and an associated rate law, $h_i(y, c_i)$, where $Y = (Y_1, \ldots, Y_n)$ presents the current state of the system. Moreover, $h_i(y, c_i)$ is stated by reaction R_i, respectively. Actually, $h_i(Y_i, \ldots, Y_n)$ generates the number of distinct combinations of R_i reactant molecules in the system when there are exactly Y_i molecules (i = 1, . . ., N). For the system in (6.3), the following functions can be shown:

$$\begin{aligned} h_1(Y, c_1) &= Y_1 Y_2\\ h_2(Y, c_2) &= \frac{Y_1(Y_1-1)}{2}\\ h_3(Y, c_3) &= Y_1 \end{aligned} \tag{6.5}$$

Also, the propensity, i.e., the hazard function, $a_j(y)$is found by $a_j(y) = c_j h_j(y)$. Thus, the equality can be achieved $a_j(y) = c_j h_j(y)$. Then, $a_j(y)$ gives the probability that an R_i reaction will occur in the time interval (t, $t + dt$) [1, 5, 8]. That is,

$$\begin{aligned} &a_j dt \triangleq \text{the probability given}\\ &Y(t) = y \text{ that } R_i \text{ reaction will occur somewhere inside}\\ &\Omega \text{ in the next infinitesimal time interval } (t, t+dt]. \end{aligned} \tag{6.6}$$

From eq. (6.6), it can be obtained through the following probability:

$$P(y, t|y_0, t_0) \triangleq \text{Prob}\{Y(t) = y, \ \ \text{given } Y(t_0) = y_0\} \tag{6.7}$$

By using eqs. (6.6) and (6.7), the following derivation is obtained:

$$\frac{\partial P(y,t|y_0,t_0)}{\partial t} = \sum_{j=1}^{M}\left[a_j(y-v_j)P(yx-v_j,t|y_0,t_0) - a_j(y)P(y,t|y_0,t_0)\right] \tag{6.8}$$

In other words, it is a CME that indicates the probability that each species will have a specified molecular population at a given future [3].

Equation (6.8) can occasionally be solved for the probability density function $Y(t)$. Here, as a novelty, a new probability function $p(\tau, j|y, t)$ is identified, rather than using the function $P(Y,t|Y_0,t_0)$ to simulate the trajectories of $Y(t)$ versus t. The associated definition can be presented as follows.

$p(\tau,j|y,t)d\tau \triangleq$ is the probability, given $Y(t) = y$ that the next reaction in the system will occur in the infinitesimal time interval $[t+\tau, t+\tau+dt]$, and will be an R_j reaction.

Then,

$$p(\tau,j|y,t) = a_j(y)\exp(-a_0(y)\tau) \tag{6.9}$$

$$a_0(y) \triangleq \sum_{j'=1}^{M} a_j(y) \tag{6.10}$$

Truthfully, eq. (6.9) is the basic mathematical equation for the SSA [3].

The time step τ should be chosen small enough that there is no marked change in the propensity function during the time interval $[t, t+\tau]$. In fact, the value of the function will not be macroscopically changed by the noninfinitely small. Essentially, the propensity function will remain stable at the value of $h_j(Y)$ for each reaction channel R_j during the time interval $[t, t+\tau]$ for $j = 1, \ldots, r$, which means that the probability of the R_j channel occurring during any infinitesimal interval dt in $[t, t+\tau]$ is $h_j(Y)dt$. Later, in the history axis of the system, the process of leaping down from the Y state at time t by the amount of τ can be described as follows. The number of times reaction channel R_j fires in $[t, t+\tau]$ is denoted by the value of k_j. A sample value of k_j can be produced from a random variable using some probability distribution such as the Poisson distribution or gamma distribution. Then, λ presents the net change in the state of the system in $[t, t+\tau]$. λ can be defined by $\lambda = \sum_{j=1}^{r} k_j v_j$, where v_j is the jth row of the stoichiometry matrix V. Therefore, the leap condition ensures that the following subtraction $|a_j(Y+\lambda) - a_j(Y)|$ is definitely infinitesimal [1, 2, 9]. In a word, the leap condition can be demonstrated by

$$|a_j(Y+\lambda) - a_j(Y)| \leq \varepsilon\, a_0(Y)$$

In this chapter, the aim is to construct confidence intervals for the population parameters k, which is the number of reactions in the system, and the time step τ in the two approximate SSAs, namely, the Poisson tau-leap, which is the most recent approximate SSAs and its extension, which is called the approximate Gillespie method, by using second- and third-order truncated Taylor expansions for the leap condition. In the current literature, the k and τ in these simulation algorithms have been used with a conservative one-sided confidence interval without controlling the significance level α. Hence, this chapter proposes realistic and accurate confidence intervals for both parameters by controlling α and by using second- and third-order truncated Taylor expansions for the leap condition so that narrower and more accurate confidence intervals can be acquired theoretically. Therefore, in this chapter, approximate SSAs will first be defined. Then, the confidence intervals inserted into the leap condition derived from the extension of second- and third-order truncated Taylor expansions will be presented. Finally, the chapter concludes with the findings illustrated by a simple example [10].

6.2 Approximate stochastic simulation algorithms

Approximate SSAs are mainly based on the Markov process, which is a method of time discretization [8]. These algorithms can be defined as effective numerical methods for the stochastic chemical kinetics [11]. Essentially, they operate under the assumption of the leap condition which is a fundamental concept in the generation of approximate methods. This means there will be no significant change in the propensity function within the chosen time interval τ.

Accordingly, the fundamental approximate SSAs can be listed as the Poisson τ-leap method, which is one of the most basic approaches, the Langevin τ-leap method, estimated midpoint techniques, binomial τ-leap method, modified Poisson τ-leap method, the new τ-selection procedure and the approximate Gillespie algorithm.

6.2.1 Poisson τ-leap method

This method aims to maximize the intervals between the selected time range τ and the selection times under the leap condition [1, 12]. In the time interval $[t, t+\tau]$ for each reaction channel R_j, a random value k_j is generated from a Poisson distribution by $\text{Poi}(h_j(Y)\tau)$, where $Y(t) = Y$ is the state vector. After substituting the boundary for the gap between

$$|h_j(Y + \lambda(Y)) - h_j(Y)|$$

where $\lambda(Y) = \sum_{j=1}^{r} k_j v_j \lambda(Y)$ infers the net change in the state of the system in $[t, t+\tau]$, a desirable value of τ can be derived. It is known that the mean of k_j equals $h_j(Y)\tau$, i.e., $E(k_j) = h_j(Y)\tau$ under $k_j \sim \text{Poi}(h_j(Y)\tau)$ and

$$\bar{\lambda}(Y, \tau) = \sum_{j=1}^{r} |h_j(Y)\tau| v_j = \tau\xi(Y) \tag{6.11}$$

That generates the expected net change in the state for the given time interval.

In this equation, j stands for the stoichiometric coefficients of the reaction j corresponding to the jth row of the net effect matrix V and $h_j(Y)$ corresponds to the hazard function of the jth reaction, which is obtained by the product of the rate constant c_j and the distinct molecular reactant combination of the underlying reaction. Afterward, $\xi(Y) = \sum_{j=1}^{r} h_j v_j$ can be demonstrated as the mean or expected state change in a unit of time by an n-dimensional vector where each ith component, $\xi_j(Y)$, represents the expected change of the ith species in a unit of time. Subsequently, the inequality is attained as

$$\left| h_j(Y + \bar{\lambda})(Y, \tau) - h_j(Y) \right| \leq \varepsilon h_0(Y) \tag{6.12}$$

by using $\lambda(Y, \tau)$ in eq. (6.11). Equation (6.12) implies that there is a restriction on the expected alterations in the hazard functions over time τ by the sum of all hazard functions $h_0 = \sum_{j=1}^{r} h_j(Y)$ and a parameter ε, which is an error control parameter lying $0 < \varepsilon < 1$. Actually, this representation gives the leap condition.

Then, the following equality is attained with the help of the first-order Taylor series expansion for the left-hand side of eq. (6.12):

$$\left| h_j(Y + \bar{\lambda}(Y, \tau)) - h_j(Y) \right| \approx \bar{\lambda}(Y, \tau) h_j(Y) = \sum_{i=1}^{n} \tau \xi_j(Y) \frac{\partial h_j(Y)}{\partial Y_i}. \tag{6.13}$$

Then, setting $b_{ji}(Y) = \frac{\partial h_j(Y)}{\partial Y_i}$ ($i = 1, \ldots, n$; $j = 1, \ldots, r$), the following inequality can be shown:

$$\tau \left| \sum_{i=1}^{n} \xi_j(Y) b_{ji}(Y) \leq \varepsilon h_0(Y) \right|$$

Eventually, the largest value of τ satisfying the leap condition for the given Y and the preselected ε is calculated via

$$\tau = \min \left\{ \frac{\varepsilon h_0(Y)}{\left| \sum_{i=1}^{n} \xi_j(Y) b_{ji}(Y) \right|} \right\} \tag{6.14}$$

Herewith, when compared to the results of the exact SSA, eq. (6.14) is not desirable because of the leap size. When $\tau \leq \frac{1}{h_0(Y)}$, the value of τ in eq. (6.14) would not be valuable, though it would be the accurate result. Hereby, the selected time step is the same as SSA. Regardless of the computational cost, the time interval in the Poisson τ-leap method is more convenient than the SSA time. Basically, there is an incremental difference between them.

Finally, replacing t by $t := t + \tau$ provides an update of the current state in the Poisson τ-leap approach. Additionally, for the value of Y, the largest value of τ should be chosen so that the leap condition is satisfied.

Alongside this, there can be the problem of negative molecular populations in the Poisson τ-leap method. To cope with this problem, some solutions can be suggested. The

Binomial τ-leap algorithm is the one of the well-known ways. It deals with this problem though this approach does not generate accurate output [11–15].

6.2.2 Langevin τ-leap method

Langevin τ-leap approach is a version of Poisson τ-leap algorithm. Because it is possible to write the Poisson distribution as a normal distribution with the same mean and variance when the number of molecules is large under the following itemized conditions. In detail, there should be no significance alteration in the hazard functions $h_j(Y)$'s ($j = 1, \ldots, r$), by the expected number of execution of reactions in the time step τ that satisfies the leap condition under the history axis of the system. Accordingly, $k_j \sim \mathrm{Poi}(h_j(Y)\tau) \approx N(h_j(Y)\tau, h_j(Y)\tau)$, where k_j is the number of times of execution of the jth reaction for $j = 1, \ldots, r$. Therefore, the steps of the Langevin τ-leap approach are as follows:

1. Choose τ such that $\tau \gg \max\{\frac{1}{h_j(Y)}\}$ under the leap condition.
2. Choose a sample value k_j by embedding $k_j = h_j(Y)\tau + (h_j(Y)\tau)^{1/2} l_j$, where a sample value l_j is generated from a normal distribution with mean zero and unit variance, $l_j \sim N(0,1)$ for each j ($j = 1, \ldots, r$).
3. Evaluate the net change in the state $\lambda(Y)$ by $\lambda(Y) = \sum_{j=1}^{r} k_j v_j$ where v_j is the n-dimensional net effect vector of the jth reaction in n-species in the system.
4. Update the system and replace the values of Y and t by $Y := Y + \lambda(Y)$ and $t := t + \tau$, respectively.

Additionally, this approach improves the speed of the Poisson τ-leap method and that the computational gain can be practical when the system is large enough to deduce the normal distribution in place of Poisson density while distinguishing the change in states. Moreover, the outputs cannot be negative values since the leap condition is in the absolute value. Hence, this algorithm can be suitable for very large systems with a large number of genes, molecules or any species in the system under long-run simulation [1, 10, 12].

6.2.3 Estimated midpoint techniques

Under the leap condition, it is necessary that the selected time step value of τ should not alter the propensity function. However, if the chosen τ which is faster in simulation with respect to SSA introduces some changes in the propensity functions, then it can encounter the problem of increased computational errors.

When solving $(dY/dt) = f(y)$ with the Euler method, which is a discretized version of the ODE, we can solve the same problem in numerical calculations as follows; Δt is the modification in line along the axis t. So $Y(t + \Delta t) = Y(t) + f(Y(t))\Delta t$ can produce calculation errors when there is a variation in f during Δt. The application of the estimated-midpoint (or second-order Runge-Kutta) approach is one of the ways to reduce the errors. In this approach, firstly, the alteration in Y is calculated by $\Delta_1 Y = f(Y(t))\Delta t$ and then it is substituted into the value of $Y(t+\Delta t)$, where $Y(t+\Delta t) = Y(t) + f(Y + \frac{\Delta_1 Y}{2} Y)\Delta t$. Later, the plausible increase in Y can be calculated by considering the slope function f at the estimated midpoint of Y during $[t, t+\tau]$ by calculating from the Euler method.

To acquire the expected state change $\bar{\lambda}(Y,\tau)$, this procedure can be adapted to the Poisson τ-leap approach. In this way, $Y + \left[\bar{\lambda}(Y,\tau)\right]$, where $[z]$ describes the largest integer in z, which corresponds to the estimated midpoint state during the leap τ.

There are two kinds of algorithms for the estimated midpoint techniques [1, 12].

6.2.3.1 Estimated midpoint τ-leap method

The first step is to calculate the expected state change $\lambda(\bar{Y},\tau)$ by $\lambda(\bar{Y},\tau) = \tau\sum_{j=1}^{r} h_j(Y)v_j$ during $[t, t+\tau]$ under the leap condition, where $h_j(Y)$ and v_j subtends the hazard function and the n-dimensional stoichiometry vector of the jth reaction, respectively. Later, the expected new state $Y' = Y + [\lambda(\bar{Y},\tau)/2]$ is evaluated, and the number of execution time k_j is produced from a Poisson distribution with mean $h_j(Y')\tau$ for each reaction $j = 1, \ldots, r$, i.e., $k_j \sim \text{Poi}(h_j(Y')\tau)$. Eventually, the net change in the state λ is acquired by $\lambda(Y) = \sum_{j=1}^{r} k_j v_j$ and the system update is done by replacing Y and t with $Y + \lambda(Y)$ and $t + \tau$, respectively [1, 12].

6.2.3.2 Estimated midpoint Langevin τ-leap method

Poisson distribution can be approached by the normal distribution under the leap condition and $\tau >> \max\{\frac{1}{h_j(Y)}\}$. Then, it can be used as the estimated midpoint Langevin τ-leap in replacement of the estimated midpoint τ-leap method in Section 2.3. In this method, with the satisfying conditions, firstly, the expected state change is computed as $\lambda(\bar{Y},\tau) = \tau\sum_{j=1}^{r} h_j(Y)v_j$. Later, a sample value $l\,j$ is generated from normal distribution with zero mean and unit variance, $l_j \sim N(0,1)$, and $k_j = h_j(Y')\tau + \left(h_j(Y')\tau\right)^{1/2} l_j$ for each $j = 1, \ldots, M$ is calculated. The rest of the steps of the algorithm is identical to the previous one. Specifically, the new expected state $Y' = Y + [\lambda(\bar{Y},\tau)/2]$ and the net change in the state $\lambda(Y) = \sum_{j=1}^{r} k_j v_j$ are calculated. As a final step, the system update is done by replacing Y and t with $Y + \lambda(Y)$ and $t + \tau\tau$, respectively.

6.2.3.3 The weak θ-midpoint trapezoidal method

In addition to these, Anderson and Mattingly [16] suggested a weak second-order method for the CLE (chemical Langevin equation). Hereby, CLE is

$$dY_t = \sum_{j=1}^{n} v_j a_j(Y_t)dt + \sum_{j=1}^{n} v_j \sqrt{a_j Y_t} dW_t^j$$

where W_t^j are independent Wiener processes and Y_i is the concentration of the molecules. This suggested approach can be described briefly as follows [16, 17]: Let $\{\mu_{1k}^i, \mu_{2k}^i : j \in (1,\ldots,n), i \in \Re\}$ be the collection of mutually independent Gaussian random variables with mean 0 and variance 1. For fixed $\theta \in (0,1)$,

$$\alpha_1 = \frac{1}{2(1-\theta)\theta} \quad \text{and} \quad \alpha_2 = \frac{(1-\theta)^2 + \theta^2}{2(1-\theta)\theta}$$

Then, for a discretization leap τ and for each $i \in (1,2,3,\ldots)$ repeat the following steps where θ-midpoint Y^* is computed at first, then the new value Y_i is found in the second stage. That is,

$$Y^* = Y_{i-1} + \sum_{j=1}^{n} v_j a_j(Y_n)\theta\tau + \sum_{j=1}^{n} v_j \sqrt{a_j(Y_n)} \eta_{1j}^{(i)} \sqrt{\theta\tau}$$

$$Y_{n+1} = Y^* + \sum_{j=1}^{n} v_j(\alpha_1 a_j(Y^*) - \alpha_2 a_j(Y_n))(1-\theta)\tau$$

$$+ \sum_{j=1}^{n} v_j \sqrt{(\alpha_1 a_j(Y^*) - \alpha_2 a_j(Y_n)) + \eta_{2j}^{(i)}} \sqrt{(1-\theta)\tau}$$

in which $y^+ = \max(0, y)$. For more information, see the study in [16].

6.2.4 Binomial τ-leap method

After the application of some τ-leap approaches, it is possible that some populations have the negative problems. As stated previously, this problem can occur as a result of two different ways. The first one is that the sample value for the reaction number can be greater than one of the molecular numbers in that reaction channel. The second one is that different reaction channels can arise simultaneously [11]. To address this issue, binomial τ-leap approaches are propounded by Chatterjee et al. [18] and Tian and Burrage [11]. According to Chatterjee et al. [18], the conditions for using the binomial τ-leap method are the same as for other τ-leap methods. That is, under the leap condition the system's history axis is used to create an approximated simulation. In the time interval $[t, t+\tau]$ given the state $Y(t)$, a maximum number of firings $k_{\max}^j$ of each reaction R_j is computed by

$$k_{\max}^j = \min_{v_{ij}<0} \left[\frac{Y_i}{|v_{ij}|} \right].$$

Herewith, the current state vector containing the number of molecules $Y_i(t)$ of all substrates $(i = 1, \ldots, n)$ at time t is denoted by Y, v_{ij} denotes each entry of the net effect matrix for the ith species and the jth reaction ($j = 1, \ldots, r$) and further $[z]$ corresponds to the greatest integer in z. Furthermore, the hazard functions h'_j s would be nearly constant during $[t, t+\tau]$ where $h_j(Y) = h_j(Y(t))$ is based on $y(t)$ [18] as described previously. An upper bound is set on the number of S_i molecules in the time interval $[t, t+\tau)$ as a precaution to eliminate negativity in the binomial τ-leap method [18]. Each R_j firing has a success probability $p = \frac{h_j(Y)\tau}{k^j_{\max}}$ (and a fail probability of $1-p$), where $\tau = \frac{f}{\sum_{j=1}^{r} h_j(Y)}$ and f presents a coarse forming factor greater than 1 under the leap condition with k_j's from binomial distribution and $k^j_{\max}$ maximum firings. If $f = 1$, then this indicates the average time increment of the exact SSA. The time and accuracy of the approximation can be determined when $f < 10^3$ or $f > 10^4$ is selected for small steps or large steps, respectively [12].

Furthermore, according to Chatterjee et al. [18], the probability that the number of firings k_j is denoted P_j belongs to the binomial distribution can be presented mathematically as follows:

$$P_{BD}\left(k_j; p, k^{(j)}_{\max}\right) = \frac{k^{(j)}_{\max}!}{k_j!\left(k^{(j)}_{\max} - k_j\right)!} p^{k_j}(1-p)^{k^{(j)}_{\max} - k_j} \tag{6.15}$$

Briefly, the steps of the binomial τ-leap algorithms can be demonstrated in the following [18]:

1. Acquire the stoichiometric coefficients v_{ij}, initial population size $Y(0)$, and the rate constants c to be necessary for calculating the propensities.
2. Start the time at $t := t_0$.
3. Repeat steps 4 to 6 to obtain the maximum time $t_{\max}$.
4. Compute the hazard function $h_j(Y(t))$ with $Y(t)$the given set $\bar{Y} := Y$.
5. Select the applicable τ and update the time $t := t + \tau$.
6. For $j = 1$ to r reactions:
   a. Catch $k^{(j)}_{\max} = \min_{v_{ij}<0}\left[\frac{\bar{Y}_i}{|v_{ij}|}\right]$, where $[z]$ is the greatest integer function.
   b. Sample k_j from eq. (6.15) as $p = \frac{h_j\tau}{k^j_{\max}}$.
   c. Set $\tilde{Y}_i := \tilde{Y}_i + v_{ij}k_j$ for $i = 1, \ldots, n$ if $v_{ij} < 0$.
   d. Go to step 6a to compute.
7. Update $Y_i := Y_i + v_{ij}k_j$ for $i = 1, \ldots, n$.
8. Return to step 3.

To sum up, this method is a way to cope with the problem of negative populations in the unbounded Poisson τ-leap method. The sample value range in the Poisson distribution is from zero to infinity. Nevertheless, unlike the Poisson approximation, the range of the sample value for Bin(n, p) is an integer from 0 to n.

Alongside the study of [18], the binomial τ-leap method is proposed by Tian and Burrage [11] to get rid of the negative population problems in the Poisson τ-leap algorithm. Over this method, k_j is generated from the binomial distribution $\mathrm{Bin}\left(N_j, \frac{h_j(Y)\tau}{N_j}\right)$ under the following condition:

$$0 \le \frac{h_j(Y)\tau}{N_j} \le 1 \tag{6.16}$$

By this means, N_j is described as follows:

1. The first-order reaction

$$S_1 \xrightarrow{c_1} S_3, h_j(Y) = c_1 Y_1, N_j = Y_1$$

2. The second-order reaction

$$S_1 + S_2 \xrightarrow{c_2} S_4, h_j(Y) = c_2 Y_1 Y_2, N_j = \min\{Y_1, Y_2\}$$

3. The homodimer formation ($Y_1 \ge 2$)

$$S_1 + S_1 \xrightarrow{c_3)} S_5, h_j(Y) = \frac{1}{2} c_3 Y_1 (Y_1 - 1), N_j = \min\left[\frac{1}{2} Y_1\right]$$

In the above, c_j is the jth reaction rate and Y_j ($i = 0, \ldots, n$) indicates the number of molecules of each species S_i. Compared to the earlier approach, which handles the negativity problem of the Poisson distribution, there is only one response. However, in this method, there are two reactions occurring simultaneously under the condition

$$0 \le \frac{h_j(Y) + h_l(Y)}{N_j} \le 1 \tag{6.17}$$

where $h_j(Y)$ and $h_l(Y)$ correspond to the propensity functions of reaction channels R_j and R_l, respectively. In other words, $h_j(Y) = N_j \frac{h_j(Y)}{N_j}$ and $h_l(Y) = N_l \frac{h_l(Y)}{N_l}$.

The sampling technique for the reaction number of channels R_j and R_l is given as follows:

1. A sample value k_{jk} for the total reaction number of R_j and R_k is produced from a binomial distribution $\mathrm{Bin}\left(N_i, \frac{h_j(Y) + h_l(Y)}{N_i} \tau\right)$, where $N_i = \min\{N_j, N_l\} \neq 0$.
2. A sample value k_j for the number of reaction R_j is produced from a binomial distribution $\mathrm{Bin}\left(k_{jl}, \frac{h_j(Y)}{h_j(Y) + h_l(Y)}\right)$.
3. The reaction number of R_l is attained from $k_l = k_{jl} - k_j$ [11, 12].

Now, the method can be identified as follows. Firstly, a step size τ selected from the τ-selection process in eq. (6.23), satisfying eq. (6.16) for each reaction channel and for a given error control parameter ε. Then, a sample value k_j is produced from the binomial distribution $\text{Bin}\left(N_j, \frac{h_j(Y)\tau}{N_j}\right)$ for $j = 1, \ldots, r$. When the reactant species are exposed to two or more reaction channels, the simultaneous reaction step size in eq. (6.17) is applied, and a sampling technique is used for these reaction channels. Finally, the system is updated by [11]

$$Y(t+\tau) = Y(t) + \sum_{j=1}^{M} v_j k_j$$

Consequently, both binomial τ-leap methods [11, 18] make allowances for resolving the problem of negativity in the Poisson τ-leap method and enhancing the computational efficiency of the algorithm due to the larger step sizes.

6.2.5 Modified Poisson τ-leap method

The modified Poisson τ-leap approach is an advanced version of the Poisson τ-leap method. The aim of this method is to prevent negative populations and achieve more accurate results than the basic Poisson τ-leap method. This approach does not allow any time step τ to generate negative populations. Smaller values of τ reduced by a factor are taken until a negative population is avoided. The algorithm is stated in the following steps.

1. Evaluate all the propensity functions and their sums:

   $$h_0(Y) = \sum_{j=1}^{r} h_j(Y)$$

   in state Y at time t.
2. Define the set of critical reactions R_j in each τ such that the current reaction channels R_j for which $h_j(Y) > 0$ and $L_j \leq n_c(\, j = 1, \ldots, r)$. Here, $h_j(Y)$ presents the hazard function for the jth reaction and $L_j = \min_{v_{ij} < 0}\left[\frac{Y_i}{|v_{ij}|}\right]$ as emphasized in the binomial τ-leap method. v_{ij} denotes the number of molecules $Y_i(t)$ of all substrates at time t and n_c shows the critical value setting the performance of the method anywhere between the original Poisson τ-leap method ($n_c = 0$) and the exact SSA ($n_c = \infty$).
3. Calculate the largest time step τ' regarding eq. (6.23) by using only the noncritical reactions j'. If there are critical reactions, then τ' will be replaced by ∞, i.e., $\tau' = \infty$.
4. If the value of τ' chosen in step 2 is less than a small multiple, like 10, of 1 $h_0(Y)$, then the leap is refused, and the exact SSA is used to obtain an accessible number of times, like 100, before again trying a τ-leap method. In addition to this, if τ' is

larger than the selected small multiple of 1 $h_0(Y)$, then it will be accepted and continued from step 5. Otherwise, stop and return to the initial step.

5. Evaluate the sum of the hazards of the critical reactions $h_0^c(Y)$. Generate τ'', $\tau'' \sim Exp\left(h_0^c(Y)\right)$ with mean $\frac{1}{h_0^c(Y)}$.
6. a. If $\tau' < \tau''$: Take $\tau = \tau'$. Set $k_j = 0$ for all the critical reactions k_j. On the other hand, generate $k_j \sim \text{Poi}\left(h_j(Y)\tau\right)$ for the noncritical reactions R_j.
   b. If $\tau'' \le \tau'$: Take $\tau = \tau''$. Produce j_c as a sample of the integer random variable with probabilities $\frac{h_j(Y)}{h_0^c(Y)}$, where j presents the index of critical reactions on j. Then, set $k_{jc} = 1$ and set $k_j = 0$ for all the other critical reactions. Generate k_j sampled from a Poisson distribution with mean $h_j(Y)\tau$ for the noncritical reactions R_j.
7. Update the time t and the state Y by $t := t + \tau$ and $Y := Y + \sum_{j=1}^{r} k_j v_{ij}$, respectively.
8. Undo step 7 if any component of Y is negative, replace τ' by $\frac{\tau'}{2}$ and go back to step 6. Otherwise, stop and return to the initial step.
9. Collect (t, Y) if desired. Return to step 1, or stop otherwise [1, 12, 13, 15].

6.2.6 The new τ-selection procedure

Thanks to the methods presented earlier, Poisson τ-leap (Section 2.1) and estimated midpoint (Section 2.3), the following inequality should be ensured:

$$h_j(Y + \lambda(Y)) - h_j(Y) \le \varepsilon h_0(Y) \tag{6.18}$$

where $h_j(Y)$'s ($j = 1, \ldots, r$) represents the hazard functions of each reaction j and $h_0(Y)$ is the sum $h_0(Y) = \sum_{j=1}^{r} h_j(Y)$. Here, $\lambda(Y)$ indicates the net change of state Y and ε shows the error control parameter ($0 < \varepsilon < 1$). Also, the largest value of τ is obtained as

$$\tau = \min_{j \in [1, r]} \left\{ \frac{\varepsilon h_0(Y)}{\left| \sum_{i=1}^{n} \xi_i(Y) b_{ji} \right|} \right\}$$

However, there is a condition to confirm this value. That is, if $\frac{1}{h_0(Y)}$ is less than a few factors, then it is better to abandon the leap and use the SSA. In this method [15], the proposal is to replace the $h_0(Y)$ term on the right in eq. (6.14) by $h_j(Y)$ to a more reasonable leap condition. In this instance, there can be a problem when $h_j(Y)$ goes to zero. A new τ-selection procedure allows for deciding the largest τ value while extending accuracy and without decelerating the simulation.

Under the leap condition, the calculation of the total change is as follows:

$$f_{jj'}(Y) = \sum_{i=1}^{n} \frac{\partial h_j(Y)}{\partial Y_i} v_{ij'} \quad (j, j' = 1, \ldots, r) \tag{6.19}$$

In eq. (6.19), the function $f_{jj'}$ denotes the hazard of reaction j as a result of the occurrence of reaction j'. Here, n represents the total number of substrates and $v_{ij'}$ denotes the net change of substrate i ($i = 1, \ldots, n$) as a consequence of the jth reaction. Additionally, v_j or $v_{j'}$ denotes an n-dimensional vector. As previously found in eq. (6.13) in Section 2.1, after the application of a first-order Taylor expansion, the following equality is obtained:

$$\begin{aligned}\Delta h_j(Y) &= h_j(Y + \lambda(\bar{Y}, \tau)) - h_j(Y) \\ &\approx \sum_{i=1}^{n} \tau\left(\frac{\partial}{\partial Y_i} h_j(Y)\right) \sum_{j'=1}^{r} h_{j'}(Y) v_{ij'},\end{aligned} \tag{6.20}$$

where $\lambda(\bar{Y}, \tau)$ is the expected change in state at time step τ. Then, as k_j is produced from a Poisson random variable with parameter $h_j(Y)\tau$, i.e., $k_j \sim \text{Poi}(h_j(Y)\tau)$, and $\lambda(\bar{Y}, \tau) = \tau \sum_{j=1}^{r} h_j(Y) v_j$, in which k_j indicates the number of occurrences of the jth reaction, the following approximation can be obtained:

$$\lambda(\bar{Y}, \tau) \approx \sum_{j=1}^{r} \text{Poi}(h_j(Y)\tau) V_j \tag{6.21}$$

By placing eq. (6.20) into eq. (6.21), substituting the two summations and using the approximation of $\lambda(\bar{Y}, \tau)$, $\Delta h_j(Y)$is obtained as follows:

$$\Delta h_j(Y) \approx \sum_{j''=1}^{r} f_{jj'}(Y) \text{Poi}(h_{j'}(Y)\tau) \tag{6.22}$$

Here, the mean and variance of $\Delta h_j(Y)$ can be represented by $\sum_{j'=1}^{r} f_{jj'}(Y) \times E(\text{Poi}(h_{j''}(Y)\tau))$ and $\sum_{j'=1}^{r} f_{jj'}(Y) \times \text{Var}(\text{Poi}(h_{j''}(Y)\tau))$, in order. Using these expressions, the following derivations can be derived:

$$E(\Delta h_j(Y)) \approx \sum_{j''=1}^{r} f_{jj'}(Y) h_j(Y)\tau \equiv \mu_j(Y)\tau$$

and

$$\text{Var}(\Delta h_j(Y)) \approx \sum_{j''=1}^{r} f_{jj'}^2(Y) h_j(Y)\tau \equiv \sigma_j^2(Y)\tau$$

for $j = 1, \ldots, r$.

Eventually, the largest appropriate τ is obtained as

$$\tau = \min_j \left\{ \frac{\varepsilon h_0(Y)}{\left|\mu_j(Y)\right|}, \frac{\varepsilon^2 h_0^2(Y)}{\sigma_j^2(Y)} \right\} \tag{6.23}$$

since $\left|\mu_j(Y)\tau\right| \leq \varepsilon h_0(Y)$ and $\sqrt{\sigma_j(Y)\tau} \leq \varepsilon h_0(Y)$. Hence,

$$\Delta h_j(Y) \approx E(\Delta h_j(Y)) \pm \sqrt{\text{Var}(\Delta h_j(Y))} \tag{6.24}$$

As the last step, similar to the other algorithms, update the system by replacing Y and t with $Y + \lambda(Y)$ and $t + \tau$ in order.

Unlike the previous Poisson τ-leap method in eq. (6.14), this new method ensures an additional state to determine the value of τ. However, it is observed that there is no situation where one is more correct than the other. When using one of them, the computationally simpler one can be chosen [10, 19, 12].

6.2.7 Approximate Gillespie algorithm

The Gillespie algorithm is one of the alternative methods of the Poisson τ-leap approach. It essentially belongs to the extension of the exact Gillespie method [12]. Here, k number of reactions is produced from the gamma distribution $\Gamma(\cdot)$ with a parameter $\sum_{j=1}^r h_j(Y)$, where each of them occurs in an exponential time step t, rather than a single reaction at a time. Accordingly, it can be shown that $\tau \sim \Gamma(k, h_0(Y))$, where τ denotes the time interval of k reactions in the total hazard, $h_0(Y), h_0(Y) = \sum_{j=1}^r h_j(Y)$. In this way, replacing t by $t := t + \tau$ and changing the current state Y by $Y := Y + \lambda(Y)$, where the net change in the state is found via $\lambda(Y) = \sum_{j=1}^r k_j v_j$, the system is updated. In this statement v_j indicates the net effect of the jth reaction by presenting the jth row of the net effect matrix V as used previously. In this way, it is assumed that the essential time for every reaction corresponds to that of Gillespie. Hence, the total number of reactions during the interval τ is specified by checking k in each time interval under the assumptions. To this end, k is initially defined by satisfying the leap condition in each time step. Later, the difference in the hazard function $\Delta h_j(Y)$ $(j = 1, \ldots, r)$ is approximated by the first-order Taylor expansion in the time interval $[t, t + \tau]$ in such a way that the following equality can be found as performed in the Poisson τ-leap approach:

$$\Delta h_j(Y) = h_j(Y + \lambda(\bar{Y}, \tau)) - h_j(Y) \approx \lambda(\bar{Y}, \tau) h_j(Y) = \sum_{i=1}^n \lambda(\bar{Y}, \tau) \frac{\partial h_{ij}(Y)}{\partial Y_i} \tag{6.25}$$

in which the expected change in the state by considering k simultaneous reactions is calculated by

$$\lambda(\bar{Y}, \tau) = Y(t+\tau) - Y(t) = \sum_{j=1}^{r} k_j v_j$$

In the above statement, k_j presents the number of times of the jth reaction, and v_j shows the net effect of the jth reaction by denoting the jth row of the net effect matrix V as before. Thus, by using a gamma distribution, it can be demonstrated that $\tau \sim \Gamma(k, h_0(Y))$ where $k = E(\tau)h_0(Y)$. Here, $E(\tau)$ is the expected τ on average.

Then, by replacing this k into eq. (6.25), the approximation of

$$\Delta h_j(Y) \approx \sum_{j=1}^{r} f_{jj'}(Y)\tau h_0(Y)$$

is gained where the total change in the hazard of the reaction j' is represented in terms of $f_{jj'}$ via

$$f_{jj'} = \sum_{i=1} \frac{\partial h_j(Y)}{\partial Y_i} v_{ij}$$

for the execution of the reaction j'. Finally, in order to get the confidence interval, the following expression is written as

$$\Delta h_j(Y) \approx E(\Delta h_j(Y)) \pm \sqrt{\mathrm{Var}(\Delta h_j(Y))} \tag{6.26}$$

where

$$E(\Delta h_j(Y)) \approx \sum_{j'=1}^{r} f_{jj'}(Y)E(\tau)h_0(Y) = \sum_{j=1}^{r} f_{jj'}(Y)\frac{k}{h_0(Y)}h_0(Y) = k\sum_{j=1}^{r} f_{jj'}(Y) \tag{6.27}$$

and

$$\mathrm{Var}(\Delta h_j(Y)) \approx \sum_{j'=1}^{r} f_{jj'}^2(Y)\mathrm{Var}(\tau)h_0(Y) = \sum_{j=1}^{r} f_{jj'}^2(Y)\frac{k}{h_0(Y)}h_0(Y) = k\sum_{j=1}^{r} f_{jj'}^2(Y) \tag{6.28}$$

By substituting eqs. (6.27) and (6.28) into the required leap condition in eq. (6.12), the following statement can be derived:

$$|k|\sum_{j'=1}^{r} f_{jj'}(Y) \leq \varepsilon h_0(Y)$$

and

$$\sqrt{\left(k\sum_{j'=1}^{r} f_{jj'}^2(Y)\right)} \leq \varepsilon h_0(Y)$$

As a result, the optimal k is found from

$$k = \min_j \left\{ \frac{\varepsilon h_0(Y)}{\left|\sum_{j'=1}^{r} f_{jj'}(Y)\right|}, \frac{\varepsilon^2 h_0^2(Y)}{\left|\sum_{j'=1}^{r} f_{jj'}^2(Y)\right|} \right\} \tag{6.29}$$

Up to now, most derivatives have been found by substituting the distributional features of k and τ into the leap condition and obtaining a conservative confidence interval for the Poisson distribution [13]. However, the confidence intervals are formed as one-sided and by using a fixed significance level α which approximately sets the tabulated value to 1. Moreover, they produce large intervals that reduce the accuracy of approximate approaches. In this way, alternative confidence intervals are generated to achieve more accurate results.

6.3 Extension of leap condition

In this part, $h_0(Y)$ is achieved by using the second- and third-order truncated Taylor formulas. These expansions provide more information about the distribution, as they offer insights into the variance and covariance, respectively. In fact, these are also known as the second and third moments. Here, the confidence intervals become narrower compared to the results from previous studies [10].

6.3.1 Extension of leap condition by the second-order Taylor expansion

The second-order truncated Taylor expansion is applied rather than using the first-order in the following way:

$$\begin{aligned} h_j\left(Y + \bar{\lambda}(Y,\tau)\right) &\approx h_j(Y) + \bar{\lambda}(Y,\tau)\nabla h_j(Y) + \frac{1}{2}\bar{\lambda}(Y,\tau)H(Y)\bar{\lambda}^T(Y,\tau) \\ h_j\left(Y + \bar{\lambda}(Y,\tau)\right) - h_j(Y) &\approx \bar{\lambda}(Y,\tau)\nabla h_j(Y) + \frac{1}{2}\bar{\lambda}(Y,\tau)H(Y)\bar{\lambda}^T(Y,\tau) \\ \Delta h_j(Y) &\approx \bar{\lambda}(Y,\tau)\nabla h_j(Y) + \frac{1}{2}\bar{\lambda}(Y,\tau)H(Y)\bar{\lambda}^T(Y,\tau) \end{aligned} \tag{6.30}$$

where $\nabla h_j(Y)$ is the gradient of $h_j(Y)$ such that $\nabla h_j(Y) = \sum_{i=1}^{n} \frac{\partial h_j(Y)}{\partial Y_i}$, $H(Y)$ presents a Hessian matrix of $h_j(Y)$ such that $H(Y) = \sum_{i,l=1}^{n} \frac{\partial^2 h_j(Y)}{\partial Y_i \partial Y_l}$ and $\bar{\lambda}(Y,t) = Y(t+\tau) - Y(t) = \sum_{j=1}^{r} k_j v_j$. Then, $\Delta h_j(Y)$ can be approximated as follows:

$$
\begin{aligned}
\Delta h_j(Y) &\approx \bar{\lambda}(Y,\tau)\nabla h_j(Y) + \frac{1}{2}\bar{\lambda}(Y,\tau)H(Y)\bar{\lambda}^T(Y,\tau) \\
\Delta h_j(Y) &\approx \sum_{i=1}^{n} \bar{\lambda}_i(Y,\tau)\frac{\partial h_{ij}(Y)}{\partial Y_i} + \frac{1}{2}\sum_{i,l=1}^{n} \bar{\lambda}_i^T(Y,\tau)\frac{\partial^2 h_{ij}(Y)}{\partial Y_i \partial Y_l}\bar{\lambda}_i(Y,\tau)
\end{aligned} \tag{6.31}
$$

where $\lambda_i^-(Y,\tau) = \sum_{j=1}^{r} k_j v_{ji}$ and $\lambda_i^-\, T(Y,\tau) = \sum_{j=1}^{r} v_{ji} k_j$. Accordingly, the subsequent statement can be found:

$$
\Delta h_j(Y) \approx k\sum_{j'=1}^{r} f_{jj'}(Y) + \frac{1}{2}k^2\sum_{j'=1}^{r} g_{jj'}(Y) \tag{6.32}
$$

where

$$
f_{jj'}(Y) = \sum_{i=1}^{r} v_{ij'}\frac{\partial h_j(Y)}{\partial Y_i}
$$

and

$$
g_{jj'}(Y) = \sum_{l=1}^{n}\sum_{i=1}^{n} v_{ij'}\frac{\partial^2 h_j(Y)}{\partial Y_i \partial Y_l} v_{j'i}
$$

Then, since τ is generated from a gamma distribution with parameters k and $h_0(Y)$, i.e., $\tau \sim \Gamma(k, h_0(Y)), k$ can be written as $k = \tau \,.\, h_0(Y)$ by using the expectation. After plugging this final expression k into eq. (6.32), $\Delta h_j(Y)$ can be obtained in the following way:

$$
\Delta h_j(Y) \approx \sum_{j'=1}^{r}\left(\tau h_0(Y) f_{jj'}(Y) + \frac{h_0^2(Y)}{2}\tau^2 g_{jj'}(Y)\right) \tag{6.33}
$$

Accordingly, we get

$$
\Delta h_j(Y) \approx E(\Delta h_j(Y)) \pm z_{\alpha/2}\sqrt{\operatorname{Var}(\Delta h_j(Y))} \tag{6.34}
$$

as seen in the previous studies in Section 2.7. Then the mean of eq. (6.33), i.e., $E(\Delta h_j(Y))$, can be computed by

$$
\begin{aligned}
E(\Delta h_j(Y)) &\approx E\left(\sum_{j'=1}^{r}\left(\tau h_0(Y) f_{jj'}(Y) + \frac{h_0^2(Y)}{2}\tau^2 g_{jj'}(Y)\right)\right) \\
&= \sum_{j'=1}^{r}\left(E(\tau) h_0(Y) f_{jj'}(Y) + E(\tau^2)\frac{h_0^2(Y) g_{jj'}(Y)}{2}\right)
\end{aligned} \tag{6.35}
$$

As

$$E(\tau) = \frac{k}{h_0(Y)} \text{ and } E(\tau^2) = \frac{k(k+1)}{h_0^2(Y)}$$

$$\begin{aligned}
E(\Delta h_j(Y)) &\approx E\left(\sum_{j'=1}^{r}\left(\tau h_0(Y) f_{jj'}(Y) + \frac{h_0^2(Y)}{2}\tau^2 g_{jj'}(Y)\right)\right) \\
&= \sum_{j'=1}^{r}\left(\frac{k}{h_0(Y)} h_0(Y) f_{jj'}(Y) + \frac{k(k+1)}{h_0^2(Y)} \frac{h_0^2(Y) g_{jj'}(Y)}{2}\right) \\
&= \sum_{j'=1}^{r}\left(\frac{k^2}{2} g_{jj'}(Y) + k\left(f_{jj'}(Y) + \frac{g_{jj'}(Y)}{2}\right)\right) \\
&= \sum_{j'=1}^{r}\left(\left(k\sqrt{\frac{g_{jj'}(Y)}{2}} + \frac{\left(f_{jj'}(Y) + \frac{g_{jj'}(Y)}{2}\right)}{\sqrt{2 g_{jj'}(Y)}}\right)^2 - \left(\frac{\left(f_{jj'}(Y) + \frac{g_{jj'}(Y)}{2}\right)}{\sqrt{2 g_{jj'}(Y)}}\right)^2\right)
\end{aligned} \tag{6.36}$$

Then, by inserting this final expression in eq. (6.36) into the leap condition, similar to the earlier works [10, 14, 20], the following inequality can be attained:

$$\left|\sum_{j'=1}^{r}\left(\left(k\sqrt{\frac{g_{jj'}(Y)}{2}} + \frac{\left(f_{jj'}(Y) + \frac{g_{jj'}(Y)}{2}\right)}{\sqrt{2 g_{jj'}(Y)}}\right)^2 - \left(\frac{\left(f_{jj'}(Y) + \frac{g_{jj'}(Y)}{2}\right)}{\sqrt{2 g_{jj'}(Y)}}\right)^2\right) \le \varepsilon h_0(Y)\right| \tag{6.37}$$

After that,

$$\sum_{j'=1}^{r}\left(k\sqrt{\frac{g_{jj'}(Y)}{2}} + \frac{\left(f_{jj'}(Y) + \frac{g_{jj'}(Y)}{2}\right)}{\sqrt{2 g_{jj'}(Y)}}\right)^2 \le \varepsilon h_0(Y) + \left(\frac{\left(f_{jj'}(Y) + \frac{g_{jj'}(Y)}{2}\right)}{\sqrt{2 g_{jj'}(Y)}}\right)^2 \tag{6.38}$$

Using the inequality in (6.38), the following expressions are derived:

$$k \le \frac{\sqrt{\varepsilon h_0(Y) + \left(\frac{\left(f_{jj'}(Y) + \frac{g_{jj'}(Y)}{2}\right)}{\sqrt{2 g_{jj'}(Y)}}\right)^2} - \sum_{j'=1}^{r} \frac{\left(f_{jj'}(Y) + \frac{g_{jj'}(Y)}{2}\right)}{\sqrt{2 g_{jj'}(Y)}}}{\sum_{j=1}^{r} \sqrt{\frac{g_{jj'}(Y)}{2}}} \tag{6.39}$$

Further, by using the statement in eq. (6.38), the following inequality can be obtained:

$$\frac{\sqrt{\left(\frac{\left(f_{jj'}(Y)+\frac{g_{jj'}(Y)}{2}\right)}{\sqrt{2g_{jj'}(Y))}}\right)^2 - \varepsilon h_0(Y) - \sum_{j'=1}^{r}\frac{\left(f_{jj'}(Y)+\frac{g_{jj'}(Y)}{2}\right)}{\sqrt{2g_{jj'}(Y)}}}}{\sum_{j=1}^{r}\sqrt{\frac{g_{jj'}(Y)}{2}}} \leq k \tag{6.40}$$

In a similar way, the calculation of $E(\Delta h_j(Y)), \mathrm{Var}(\Delta h_j(Y))$ is attained by the following expressions:

$$\begin{aligned}\mathrm{Var}(\Delta h_j(Y)) &\approx \mathrm{Var}\left(\sum_{j'=1}^{r}\left(\tau h_0(Y)f_{jj'}(Y)+\frac{h_0^2(Y)}{2}\tau^2 g_{jj'}(Y)\right)\right)\\ &= \sum_{j'=1}^{r}\left(h_0^2(Y)f_{jj'}(Y)\mathrm{Var}(\tau)+\frac{h_0^4 g_{jj'}(Y)}{4}\mathrm{Var}(\tau^2)\right.\\ &\quad \left.+ h_0^3(Y)f_{jj'}(Y)g_{jj'}(Y)\mathrm{Cov}(\tau,\tau^2)\right)\end{aligned} \tag{6.41}$$

As $\tau \sim \Gamma(k, h_0(Y)$, it is known that $\mathrm{Var}(\tau) = \frac{k}{h_0^2(Y)}$, $\mathrm{Var}(\tau^2) = \frac{4k^3+10k^2+6k}{h_0^4(Y)} = \frac{2k(k+1)(2k+3)}{h_0^4(Y)}$ and $\mathrm{Cov}(\tau,\tau^2) = \frac{2k(k+1)}{h_0^3(Y)}$. Substituting these values into the equation of $\mathrm{Var}(\Delta h_j(Y))$ (eq. (6.41)), the following expression is derived:

$$\begin{aligned}\mathrm{Var}(\Delta h_j(Y)) &\approx \sum_{j'=1}^{r}\left(h_0^2(Y)f_{jj'}(Y)\frac{k}{h_0^2(Y)}+\frac{h_0^4(Y)g_{jj'}(Y)}{4}\frac{2k(k+1)(2k+3)}{h_0^4(Y)}\right.\\ &\quad \left.+ h_0^3(Y)f_{jj'}(Y)g_{jj'}(Y)\frac{2k(k+1)}{h_0^3(Y)}\right)\\ &= \sum_{j'=1}^{r}\left(f_{jj'}^2(Y)k+\frac{g_{jj'}^2(Y)2k(k+1)(2k+3)}{4}+f_{jj'}(Y)g_{jj'}(Y)2k(k+1)\right)\end{aligned} \tag{6.42}$$

After plugging eq. (6.42) into the leap condition, it is obtained that

$$k \leq \frac{\sqrt{\varepsilon^2 h_0^2(Y) - A(Y) + \sum_{j'=1}^{r} B^2(Y)} - \sum_{j'=1}^{r} B(Y)}{\sum_{j=1}^{r} g_{jj'}(Y)} \tag{6.43}$$

where $A(Y)$ and $B(Y)$ are as follows:

$$A(Y) = \sum_{j'=1}^{r}\left(2f_{jj'}(Y)g_{jj'}(Y)+f_{jj'}^2(Y)+\frac{3}{2}g_{jj'}^2(Y)\right)$$

$$B(Y) = \left(\frac{\frac{5}{2}g_{jj'}(Y)+2f_{jj'}(Y)g_{jj'}(Y)}{2g_{jj'}(Y)}\right)$$

Accordingly,

$$k = \min_{j\in[1,r]} \left[\frac{\sqrt{\varepsilon h_0(Y) + \left(\frac{\left(f_{jj'}(Y) + \frac{g_{jj'}(Y)}{2}\right)}{\sqrt{2g_{jj'}(Y)}} \right)^2} - \sum_{j=1}^{r} \frac{\left(f_{jj'}(Y) + \frac{g_{jj'}(Y)}{2}\right)}{\sqrt{2g_{jj'}(Y)}}}{\sum_{j'=1}^{r} \sqrt{\frac{g_{jj'}(Y)}{2}}} \right] \tag{6.44}$$

$$\frac{\sqrt{\varepsilon^2 h_0^2(Y) - A(Y) + \sum_{j=1}^{r} B^2(Y)} - \sum_{j'=1}^{r} B(Y)}{\sum_{j'=1}^{r} g_{jj'}(Y)}$$

where $A(Y) = \sum_{j'=1}^{r} (2f_{jj'}(Y)g_{jj'}(Y) + f_{jj'}^2(Y) + \frac{3}{2}g_{jj'}^2(Y))$ and $B(Y) = \left(\frac{\frac{5}{2}g_{jj'}(Y) + 2f_{jj'}(Y)g_{jj'}(Y)}{2g_{jj'}(Y)} \right)$. Hereby, as defined above, the function $g_{jj'}(Y)$ comes from the second-order truncated Taylor expansion. Moreover, it is seen that when the function $g_{jj'}(Y)$ becomes closer to zero, the expression can be the same as the output from the first-order Taylor expansion.

6.3.2 Extension of leap condition via the third-order truncated Taylor expansion

Similar to the second-order truncated Taylor expansion, we apply the third-order truncated Taylor expansion to approximate $\Delta h_j(Y)$ with the leap condition. Then, we have the following statements:

$$\begin{aligned} \Delta h_j(Y) \approx & \sum_{i=1}^{n} \bar{\lambda}_i(Y,\tau) \frac{\partial h_j(Y)}{\partial Y_i} + \frac{1}{2} \sum_{i,l=1}^{n} \bar{\lambda}_i^T(Y,\tau) \frac{\partial^2 h_j(Y)}{\partial Y_i \partial Y_l} \bar{\lambda}_i(Y,\tau) \\ & + \frac{1}{6} \sum_{i,l,m=1}^{n} \frac{\partial^3 h_j(Y)}{\partial Y_i \partial Y_l \partial Y_m} \bar{\lambda}_i(Y,\tau) \bar{\lambda}_l(Y,\tau) \bar{\lambda}_m(Y,\tau) \end{aligned} \tag{6.45}$$

where $\bar{\lambda}_i(Y,\tau) = \sum_{j=1}^{r} k_j v_{ji}$, $\bar{\lambda}_l(Y,\tau) = \sum_{j=1}^{r} k_j v_{jl}$ and $\bar{\lambda}_m(Y,\tau) = \sum_{j=1}^{r} k_j v_{jm}$. Accordingly, we can obtain the subsequent statement:

$$\Delta h_j(Y) \approx k \sum_{j'=1}^{r} f_{jj'}(Y) + \frac{1}{2} k^2 \sum_{j'=1}^{r} g_{jj'}(Y) + \frac{1}{6} k^3 \sum_{j'=1}^{r} l_{jj'}(Y) \tag{6.46}$$

in which

$$f_{jj'}(Y) = \sum_{i=1}^{r} v_{ij'} \frac{\partial h_j(Y)}{\partial Y_i}$$

$$g_{jj'}(Y) = \sum_{l=1}^{n} \sum_{i=1}^{n} v_{ij'} \frac{\partial^2 h_j(Y)}{\partial Y_i \partial Y_l} v_{j'i}$$

and

$$l_{jj'}(Y) = \sum_{m=1}^{n} \sum_{l=1}^{n} \sum_{i=1}^{n} \frac{\partial^3 h_j(Y)}{\partial Y_i \partial Y_l \partial Y_m} v_{j'i} v_{j'l} v_{j'm}$$

Similar to previous derivations, under the gamma distribution, it is known that $k = \tau h_0(Y)$. Subsequently,

$$\Delta h_j(Y) \approx \sum_{j'=1}^{r} \left(\tau h_0(Y) f_{jj'}(Y) + \frac{1}{2} \tau^2 h_0^2(Y) g_{jj'}(Y) + \frac{1}{6} \tau^3 h_0^3(Y) l_{jj'}(Y) \right) \tag{6.47}$$

Then, $E(\Delta h_j(Y))$ and $\mathrm{Var}(\Delta h_j(Y))$ are derived by

$$\begin{aligned} E(\Delta h_j(Y)) &\approx \sum_{j'=1}^{r} \left(E(\tau) h_0(Y) f_{jj'}(Y) + \tfrac{1}{2} E(\tau^2) h_0^2(Y) g_{jj'}(Y) + \tfrac{1}{6} E(\tau^3) h_0^3(Y) l_{jj'}(Y)\right) \\ &= \sum_{j'=1}^{r} \left(k f_{jj'}(Y) + \frac{1}{2} k(k+1) g_{jj'}(Y) + \frac{1}{6} k(k+1)(k+2) l_{jj'}(Y) \right) \end{aligned} \tag{6.48}$$

and

$$\begin{aligned} \mathrm{Var}(\Delta h_j(Y)) &\approx \sum_{j'=1}^{r} \mathrm{Var}\left(\tau h_0(Y) f_{jj'}(Y) + \frac{1}{2} \tau^2 h_0^2(Y) g_{jj'}(Y) + \frac{1}{6} \tau^3 h_0^3(Y) l_{jj'}(Y) \right) \\ &= \sum_{j'=1}^{r} \left(h_0^2(Y) f_{jj'}^2(Y) \mathrm{Var}(\tau) + \frac{h_0^4(Y) g_{jj'}^2(Y)}{4} \mathrm{Var}(\tau^2) \right. \\ &\quad + \frac{h_0^6(Y) l_{jj'}^2(Y)}{36} \mathrm{Var}(\tau^3) + h_0^3(Y) f_{jj'}(Y) g_{jj'}(Y) \mathrm{Cov}(\tau, \tau^2) \\ &\quad \left. + \frac{h_0^5(Y) g_{jj'}(Y) l_{jj'}(Y)}{6} \mathrm{Cov}(\tau^2, \tau^3) + \frac{h_0^4(Y) f_{jj'}(Y) l_{jj'}(Y)}{3} \mathrm{Cov}(\tau, \tau^3) \right) \end{aligned}$$

As $k \sim \Gamma(k, h_((0)(Y))$, it is known that

$$\mathrm{Var}(\tau^2) = \frac{6k^3 + 9k^2 + 6k - 1}{h_0^4(Y)}$$

$$\mathrm{Var}(\tau^3) = \frac{9k^5 + 72k^4 + 213k^3 + 180k^2 + 210k}{h_0^6(Y)}$$

$$\mathrm{Cov}(\tau, \tau^2) = \frac{2k(k+1)}{h_0^3(Y)}$$

$$\mathrm{Cov}(\tau^2, \tau^3) = \frac{k(k+1)(k+2)(6k+12)}{h_0^5(Y)}$$

$$\mathrm{Cov}(\tau, \tau^3) = \frac{3k(k+1)(k+2)}{h_0^4(Y)}$$

Then,

$$\begin{aligned}\mathrm{Var}(\Delta h_j(Y)) \approx \sum_{j'=1}^{r} \Bigg(& f_{jj'}^2(Y)k + \frac{g_{jj'}^2(Y)}{4}(6k^3+9k^2+6k-1) \\ & + \frac{l_{jj'}^2(Y)}{36}(9k^5+72k^4+213k^3+180k^2+210k) \\ & + f_{jj'}(Y)g_{jj'}(Y)(2k(k+1)) \\ & + \frac{g_{jj'}(Y)l_{jj'}(Y)}{6}(k(k+1)(k+2)(6k+12)) \\ & + \frac{f_{jj'}(Y)l_{jj'}(Y)}{3}(3k(k+1)(k+2))\Bigg)\end{aligned} \tag{6.49}$$

Inserting the statements in eq. (1.48) into the required leap condition results in

$$k \le \frac{\sqrt{\varepsilon h_0(Y) + \sum_{j'=1}^{r}\left(\frac{3\left(g_{jj'}(Y)+l_{jj'}(Y)\right)^2}{2l_{jj'}(Y)} + f_{jj'}(Y) + \frac{l_{jj'}(Y)}{3}\right)}}{\sum_{j'=1}^{r}\sqrt{\frac{l_{jj'}(Y)}{6}}} - \frac{\left(\frac{g_{jj'}(Y)+l_{jj'}(Y)}{2}\right)\sqrt{\left(\frac{6}{l_{jj'}(Y)}\right)}}{\sum_{j'=1}^{r}\sqrt{\frac{l_{jj'}(Y)}{6}}} \tag{6.50}$$

As seen above, we can obtain the new values, which produce narrower intervals for k and τ as expected. This is because, as validated theoretically, the second and third-order truncated Taylor expansions are more accurate approximations than using the first-order Taylor expansion.

6.4 Application: basic numerical example

To demonstrate the accuracy of the claim, a basic example is used here. Consider the following system:

$$
\begin{aligned}
Y_1 + Y_2 &\xrightarrow{c_1} Y_5 \\
Y_1 &\xrightarrow{c_2} Y_1 + Y_3 \\
Y_5 &\xrightarrow{c_3} Y_1 + Y_2 \\
Y_5 &\xrightarrow{c_4} Y_3 + Y_4 \\
2Y_4 &\xrightarrow{c_5} Y_1 \\
Y_1 &\xrightarrow{c_6} 2Y_4 \\
Y_3 &\xrightarrow{c_7} \varnothing \\
Y_4 &\xrightarrow{c_8} \varnothing
\end{aligned}
$$

Hereby, the state vector Y is defined of $Y_1, \ldots, Y_5$ such that $Y = [Y_1, Y_2, Y_3, Y_4, Y_5]$. Then, the vector c which consists of the reaction rate constants c_j's for each reaction channel R_j for $j = 1, \ldots, 8$, and the stoichiometry matrix V can be written as follows:

$$
V = \begin{bmatrix}
-1 & 0 & 1 & 0 & 1 & -1 & 0 & 0 \\
-1 & 0 & 1 & 0 & 0 & 0 & 0 & 0 \\
0 & 1 & 0 & 0 & 0 & 0 & -1 & 0 \\
0 & 0 & 0 & 1 & -2 & 2 & 0 & -1 \\
1 & 0 & -1 & 0 & 0 & 0 & 0 & 0
\end{bmatrix}
$$

Then, the state $Y = [1,\ 12,\ 2,\ 10,\ 0]$ and the reaction rate constant $c = [1,\ 0.5,\ 1,\ 0.05,\ 1,\ 0.25,\ 0.1,\ 0.1]$ are given. Our hazard functions for each reaction are as follows:

$$
\begin{aligned}
h_1(Y) &= c_1 Y_1 Y_2 \\
h_2(Y) &= c_2 Y_1 \\
h_3(Y) &= c_3 Y_5 \\
h_4(Y) &= c_4 Y_3 \\
h_5(Y) &= c_5 Y_1 (Y_1 - 1)/2 \\
h_6(Y) &= c_6 Y_1 \\
h_7(Y) &= c_7 Y_3 \\
h_8(Y) &= c_8 Y_4
\end{aligned}
$$

For this value, $h_0(Y)$ is computed as $h_0(Y) = \sum_{j=1}^{8} h_j(Y) = 59.5$.

When (1.44) with $\varepsilon = 10^{-2}$, the result is

$$
k_{2\text{nd}} = \min_j \lceil 0.5905, \infty \rceil \tag{6.51}
$$

So, the minimum values of k can be chosen as $k_{2nd} = 1$. To compare the previous result in the literature, k is calculated via eq. (6.29). Then, it is chosen from

$$k_{\text{AppGil}} = \min_j \lceil 11.8100,\ 139.4761 \rceil \tag{6.52}$$

Equations (6.51) and (6.52) imply that the minimum values of k can be chosen as $k_{2nd} = 1$.

Moreover, if the error control parameter ε is chosen larger, such as $\varepsilon = 10^{-1}$, then the interval becomes larger as well. For instance, in our example, the values of k become

$$k_{2nd} = \min_j \lceil 5.9050, \infty, \rceil$$

and

$$k_{\text{AppGil}} = \min_j \lceil 118, 13948 \rceil$$

So, minimum values of k can be chosen as $k_{2nd} = 6$. Instead of choosing larger ε, when ε becomes smaller, such as $\varepsilon = 10^{-3}$, then the interval is also smaller:

$$k_{2nd} = \min_j \lceil 0.0591, \infty, \rceil \tag{6.53}$$

and

$$k_{\text{AppGil}} = \min_j \lceil 1.1810, 1.3948 \rceil \tag{6.54}$$

Thus, minimum values of k can be chosen as $k_{2nd} = 0$. However, as the value of k denotes the number of reactions, 0 reaction make no sense. Therefore, this just proves that eq. (6.51) gives the smallest value of k and a narrower interval for k when it is compared with others.

This is the basic example of comparison between the results from the second-order truncated Taylor expansion under the gamma distribution and the previous output in the approximate Gillespie algorithm. To obtain more desirable results, a real system and real inputs should be used, as the given values in this example are chosen randomly. Consequently, this example demonstrates that the new value of k is smaller than or equal to the existing value of k in the literature. Furthermore, the new intervals are narrower than the intervals obtained from the approximate Gillespie algorithm. In addition, the other new outputs in this study can be compared in the same way.

6.5 Conclusion

Many chemical reactions occur in biological systems. When these are simulated by SSAs such as the Gillespie method, the next reaction method and the first-reaction method, these approaches are not computationally convenient for large systems although they produce accurate generation of systems. For this reason, approximate SSAs such as the Poisson τ-leap method and the approximate Gillespie algorithm become useful. They can simulate the system faster than SSAs which are based on the leap condition. Under this condition, the time step τ is chosen so that there should be no significant change in the hazard function during the time interval $[t, t+\tau]$ [1, 13]. In this chapter, under the leap condition, the confidence interval for the number of simultaneous reactions k in the time interval is derived. Here, we have used the second- and third-order truncated Taylor expansion of the hazard function under the gamma distribution. As a result, we have obtained narrower intervals for the value of k than previous studies. We have shown the application of our derivatives via a basic example. As future work, we consider producing simulation studies with distinct scenarios so that we can numerically evaluate the performance of the obtained theoretical outputs.

References

[1] D. T. Gillespie. Approximate accelerated stochastic simulation of chemically reacting systems. The Journal of Chemical Physics, 115(4):1716–1733, 2001.

[2] D. T. Gillespie. Simulation methods in systems biology. 125–167, 2008.

[3] D. T. Gillespie. Stochastic simulation of chemical kinetics. Annual Review of Physical Chemistry, 58(1):35–55, 2007. PMID: 17037977.

[4] D. F. Anderson. Incorporating post leap checks in tau-leaping. The Journal of Chemical Physics, 128(5):054103, 2008.

[5] D. T. Gillespie. A rigorous derivation of the chemical master equation. Physica A, 188(404):25, 1992.

[6] S. Asmussen and P. W. Glynn. Stochastic Simulation: Algorithms and Analysis, volume 57. Springer, 2007.

[7] D. T. Gillespie. Exact stochastic simulation of coupled chemical reactions. The Journal of Physical Chemistry, 81(25):2340–2361, 1977.

[8] D. J. Wilkinson. Stochastic Modelling for Systems Biology. Chapman and Hall/CRC, 2006.

[9] Y. Cao and L. Petzold. Slow-scale tau-leaping method. Computer Methods in Applied Mechanics and Engineering, 197(43–44): 3472–3479, 2008.

[10] S. Demirbüken and V. Purutçuoğlu. Extension of leap condition in approximate stochastic simulation algorithms of biological networks.

[11] T. Tian and K. Burrage. Binomial leap methods for simulating stochastic chemical kinetics. The Journal of Chemical Physics, 121(21):10356–10364, 2004.

[12] V. Purutçuoğlu and E. Wit. An approximation algorithm based on leap condition for stochastical simulation of biomedical systems. In: Proceeding of the 4th International Conference "Inverse problems: Modelling and Simulation, 2008.

[13] Y. Cao, D. T. Gillespie and L. R. Petzold. Avoiding negative populations in explicit Poisson tau-leaping. The Journal of Chemical Physics, 123(5):054104, 2005.

[14] S. Demirbüken and V. Purutçuoğlu. Extension of leap condition in approximate stochastic simulation algorithms of biological networks. In: Proceeding of the 4th International Conference on Mathematics, 288–298, 2020.

[15] D. T. Gillespie and L. R. Petzold. Improved leap-size selection for accelerated stochastic simulation. The Journal of Chemical Physics, 119(16):8229–8234, 2003.

[16] D. F. Anderson and J. C. Mattingly. A weak trapezoidal method for a class of stochastic differential equations. arXiv preprint arXiv:0906.3475, 2009.

[17] Y. Hu, T. Li and B. Min. A weak second order tau-leaping method for chemical kinetic systems. The Journal of Chemical Physics, 135(2): 024113, 2011.

[18] A. Chatterjee, D. G. Vlachos and A. K. Ma. Binomial distribution based τ-leap accelerated stochastic simulation. The Journal of Chemical Physics, 122(2): 024112, 2005.

[19] M. A. Gibson and J. Bruck. Efficient exact stochastic simulation of chemical systems with many species and many channels. The Journal of Physical Chemistry A, 104(9):1876–1889, 2000.

[20] S. Demirbüken and V. Purutçuoğlu. Extension of Leap Condition in Approximate Stochastic Simulation Algorithms of Biological Networks with 2nd and 3rd Order Taylor Expansion. In: F. Yilmaz, A. Queiruga-Dios, M. J. S. Sanchez, D. Rasteiro, V. G. Martinez and J. M. Vaquero, Editors, Mathematical Methods for Engineering Applications. Cham: Springer International Publishing, 271–290, 2022.

Aditya Singh

Chapter 7 The role of supply chain management in the construction industry

Abstract: This chapter explains the concept of supply chain management and describes it in detail to give readers a rough idea about it. It also mentions how supply chain management is helping the construction industry in the current scenario. The chapter further discusses the history and evolution of supply chain management over the years. Additionally, it shows the working process of supply chain management and the numerous advantages it provides to the construction industry. Moreover, the functions and components of supply chain management are also mentioned here. The chapter also addresses the major challenges that supply chain management faces at present along with some steps to overcome those challenges. Additionally, some graphs are plotted based on various sources to support the theme of the chapter and the book. This chapter focuses on supply chain management in a particular sector, which is the construction sector, but it is also a part of supply chains that, in a way, supports the main theme of the book.

Keywords: Supply chain management, creation era, supply chain, globalization era, construction supply chain management

7.1 Introduction

Supply chain management is defined as how the supply of goods as well as services are managed or supervised amid businesses or industries and their locations. In addition it comprises how raw materials, finished goods and work in process inventory are moved as well as stored and end-to-end fulfillment of orders from the origin point to the consumption point are supplied. Node businesses, channels as well as networks that are having their constituents connected to one another are combined in the delivery or supply of products in addition to services needed in all the supply chains by their respective end customers. Supply chain management also describes when supply chain activities are designed, planned, executed, controlled as well as monitored

Acknowledgment: This research received no specific grant from any funding agency in the public, commercial, or not-for-profit sectors.

Aditya Singh, School of Civil Engineering, Lovely Professional University, Phagwara, India,
e-mail: aditya7777777k@gmail.com, aditya.11602217@lpu.in, https://orcid.org/0000-0001-9347-5627

https://doi.org/10.1515/9783110717655-007

with a given goal to leverage global logistics, in order to create net value, measure performance on a global level, create a competitive or viable infrastructure as well as synchronize supply according to the demands. Its practice attracts profoundly the domain of operations management, marketing, industrial engineering, information technology, systems engineering, procurement as well as logistics. In addition, it tries for a multimethod, unified as well as multidisciplinary method. It also requires marketing networks to perform a crucial part in it. The present research in it is apprehensive with risk management as well as sustainability-related topics apart from other area-related topics. Supply chain resilience is a significant concept that is talked over in supply chain management. There are suggestions by people that internal integration, talent management, people dimension of supply chain management, visibility or transparency, ethical issues, talent management or human capital are one of the many topics that till now have been underemphasized in the research plans as well as programs. Supply chain management can also be said to have wide ranging activities which needed planning, controlling as well as executing the supply of products from the initial point consisting of raw materials to the intermediate state consisting of the production process to the final point consisting of the distribution of the produced product to the desired location in the most practically cost-effective way. It also includes the amalgamated planning process as well as execution process needed in order to enhance the supply of raw materials, information in addition to funds in functions that generally comprises inventory management, demand planning, production, storage as well as transportation or logistics, and sourcing. Even though it shares the same goal with supply chain engineering, it is still concentrated on a highly old-style business as well as management built methods; on the other hand, the latter is concentrated on the mathematical model built methods. There are various methods that supply chain management uses with the goal to coordinate the complete supply chain including all its parts, from the initial point of raw materials supply to delivery and distribution of finished products, in order to reduce the total cost as much as possible even when there are disputes or conflicts going on between the supply chain partners. This can be understood with an example like the sale department of a company can have conflicts with the people responsible for warehouse of that company, as the sale department might need more levels of inventory to meet the demands of all the company's customers during peak season but the people responsible for taking care of the warehouse of the company might not agree with them as to decrease their holding costs.

7.1.1 Functions of supply chain management

Supply chain management can also be referred to as a cross-functional method that comprises the ways to manage the supply of raw materials in different organizations, as well as the processing of those raw materials into desired products. Additionally, it involves the supply of these desired products to the respective customers of those organizations. When organizations or companies focus on their core capabilities and become highly flexible, they decrease their ownership of distribution channels and sources of raw materials. These functions are increasingly outsourced to subcontractors or other companies capable of performing such activities in a cost-efficient and better manner. This leads to a rise in the number of organizations responsible for meeting customer demands while decreasing managerial control over day-to-day logistics operations. When there is minimal control and a higher number of supply chain partners, the concept of supply chain management comes into existence. The purpose of supply chain management is to enhance collaboration and trust among supply chain partners to improve the speed and visibility of inventory supply.

7.1.2 Supply chain resilience

Supply chain resilience is an essential part of supply chain management, and it is described as the supply chain's capability to persevere, adapt or change when required in any situation. It is said that for a long period of time, the word "resilience" was interpreted as the same as engineering resilience, which means robustness, in the field of supply chain management, leading to the idea of persistence. The famous execution of this notion was done by gauging the time to recover as well as the time to survive the supply chain in order to identify the system's weak points. In recent years, the word "resilience" has been interpreted similarly to social-ecological resilience and ecological resilience, leading to the idea of adaptation and change, respectively. Hence, the supply chain has been understood as a social-ecological system, similar to an ecosystem like a forest, continuously adapting to the external conditions of the environment, with the assistance of social players and their capability for foresight, in addition to altering or changing itself into a fundamentally novel system. This takes it in the direction of a panarchical (related to panarchy theory) interpretation of the supply chain, embedding it into a system of systems, allowing for the examination of how the supply chain interacts with systems that do not operate at the same level, such as the Earth, society and political economy, all systems that work on different levels. For instance, all three aforementioned aspects of resilience can be discussed in the context of the obstruction that occurred in the Suez Canal in 2021, where the canal was blocked by a ship for many days. Here, perseverance means to retaliate; in this instance, it means to remove the ship swiftly in order to return to the normal state quickly, so it can operate smoothly as before. Then, adaptation means to accept that the system has come to a new normal state and all actions need to be taken accordingly, which means that all ships need to be redirected around the African cape or use another mode of transportation as a replacement. Finally, change means to question the norms of linear supply chains, globalization and outsourcing, in addition to envisioning another possibility or course of action. In this instance, it leads toward circular and local supply chains that do not require international routes for transportation purposes any more.

7.1.3 Incorporation of business processes

To enhance supply chain management, there must be a shift from managing functions separately to fully integrating activities within core supply chain processes. For example, a company's purchasing department identifies its needs and places orders accordingly, while the marketing department addresses customer demands, collaborating with various retailers and distributors to meet these needs. Effective information sharing among supply chain partners can be achieved through process assimilation, where buyers and suppliers work collectively, sharing systems, jointly developing products, and exchanging information seamlessly. Lambert and Cooper highlight that a continuous flow of information is essential for the efficient functioning of an integrated supply chain. Many companies have realized that optimizing product flows is unachievable without a process-oriented approach. Lambert and Cooper [1] identified several key supply chain processes critical for this optimization, including product development and commercialization, customer service management, returns management, customer relationship management, manufacturing flow management, demand management, supplier relationship management, and order fulfillment. These processes collectively contribute to a more efficient and responsive supply chain.

Demand management is the supply chain process where a lot of work has been done. Top companies have comparable traits, which are as follows:

- Forecasting is done at the customer level.
- To decrease the lead time, they have taken initiatives.
- Strict feedback from the customers is taken into account, and the demands of the market are considered.
- Partnerships are formed not only externally but also internally by companies.

7.1.4 The elements of supply chain management

The elements of supply chain management are an important aspect that concerned individuals need to understand before managing their respective supply chains properly. The level of management as well as the incorporation of a business process linkage is typically a function of the level and number of elements or constituents added to the linkage. Therefore, increasing the level of each element or adding additional management elements can raise the level of assimilation of the business process linkage. Literature on supply chain management, the relationship between buyer and supplier and business process reengineering indicate numerous possible elements

that managers should pay attention to while managing supply relationships. Lambert and Cooper [1] recognized the major elements, which are as follows:
- Work structure
- Culture as well as attitude
- Planning as well as control
- Power and leadership structure
- Risk as well as reward structure
- Product flow facility structure
- Methods of the management
- Structure of the organization
- Structure of the information flow facility

On the other hand, there is a need for a more cautious investigation of the current literature to gain a more complete understanding of the critical supply chain elements or divisions of the previously recognized supply chain business processes, and the type of relationship these elements might hold with customers and suppliers. Boersox and Closs [2] stated that the emphasis on collaboration signifies the synergism guiding the way toward the maximum level of mutual accomplishment. A primary level channel participant is a business that is willing to take on the responsibilities of inventory ownership or consider other monetary risks, thus comprising primary level elements. In contrast, a secondary level participant is a business that engages in network relationships by providing essential services for primary participants, thus comprising secondary level elements that can assist primary participants. Additionally, third level network participants and elements assist primary level network participants and are essential divisions of secondary level elements that might also be involved. Therefore, Lambert and Cooper's structure of supply chain elements does not guide toward any clear guidance regarding the primary or secondary level supply chain elements. There is no indication of which supply chain elements should be considered primary or secondary, how these elements should be arranged to achieve a more complete structure of the supply chain or how to evaluate the given supply chain as an integrative one.

7.1.5 Supply chain management in construction sector

The idea of supply chain management originated in the manufacturing sector, particularly among motorized manufacturers who popularized it globally. They utilized tools like logistics management and just-in-time supply systems for their processes involved in the production of their concerned items. While supply chain management has limited applicability in the construction sector, it must focus more on specific features of the supply chain rather than applying the entire supply chain management industry to the construction field. The supply chain comprises numerous organizations connected in various ways, from the initial point where suppliers obtain raw materials to the point when the finished product is received by the client. These diverse processes and activities have the ability to create value through services and products for the ultimate client.

Organizations in the construction sector offer services in addition to products, which account for roughly 80% of the overall cost of a project. The resources are frequently prepared exclusively for individual projects, with inadequate possibilities for replication in other projects. Supply chain management has the ability to assist in effectively managing services in addition to products by considering when and how they should be obtained. On the other hand, it enhances the overall cost-effectiveness, functionality and quality of a given construction project.

In other words, in the construction sector, supply chain management is a relationship developed between the contractors and the suppliers in a construction project. This integrated method is the basis for the successful completion of a construction project, where all the concerned stakeholders in that given project have a complete understanding of the involved people, logistics, and resources to ensure the delivery or handover of the completed program and project are within the given time frame and budget.

According to the Farmer Review (2016), the construction sector will remain viable indefinitely. However, there is a constant risk of the total cost of a project exceeding the initial estimate by a significant margin, and at times even surpassing the total allocated budget, especially in complicated or difficult construction projects. Additionally, there are risks of political changes or funding issues during the course of the project. Fortunately, there are numerous opportunities to ensure that project costs can be controlled. One way to address this problem is through the use of intelligent and well-versed commercial management processes, along with good governance and experience in the relevant area. Thorough commercial management is the foundation of successful programs and projects. It is highly valuable not only for keeping programs on track but also for ensuring that potential risks and delays are identified. Successful construction supply chain strategies are also closely related to this, as they begin well before the construction of the project starts. Both competition and cooperation are highly valued. Additionally, these strategies involve the concerned stakeholders of the project and are both foreseeable and progressive.

In recent times, there has been an increasing inclination in the infrastructure sector to look to other sectors or industries, such as retail, manufacturing and high tech, to see what advice supply chain management can offer for successful implementation in construction. Even though these sectors or industries are completely different, there are still numerous techniques and tools that can be implemented in infrastructure, and surprisingly, they are very similar in function.

The value chain of the construction industry is composed of numerous players and is uneven, making supply chain management one of the main constituents to ensure that the project progresses smoothly. Additionally, all concerned stakeholders must be fully aware of the performance and progress in their respective areas to ensure the project is completed within the given time frame and budget.

The archetypal supply chain in a construction project comprises engineers, material suppliers, main contractors and subcontractors, each responsible for a specific role in the project. All of them join hands to complete a given project. The intricate chain generally consists of temporary relationships and unusual procedures due to the uniqueness of each construction project. Additionally, the players involved in a construction project typically work together only once in their careers, or only a few fortunate ones are able to collaborate on more than one project.

Unfortunately, after COVID-19 appeared and spread worldwide, the supply chain management was completely disrupted. This affected the construction sector by causing a rise in the cost of materials, unavailability of materials and increased delays in construction projects, which ultimately increased the overall project costs by a significant margin. According to the report from Dodge Data, general contractors were afraid that costs would continue to increase in 2021.

From the viewpoint and understanding of CEMEX Ventures, based on data collected by their analyst team, the supply chain management presents one of the highest opportunities in the construction industry. This is because supply chain management has remained one of the most underinvested areas or has been greatly ignored in the construction sector.

The head of CEMEX Ventures, Gonzalo Galindo, felt that the recent pandemic increased the need to develop novel opportunities, technologies and solutions, as well as enhance the existing ones in order to optimize the supply chain of the construction industry. He believed that the supply chain plays an important role in the quick recovery of the construction industry after the pandemic.

Basically, it can be said that the recent pandemic greatly affected the construction industry. To make the construction industry function as it did before the pandemic or to bring it back on track through traditional methods, it will take many years to accomplish. However, with the help of construction supply chain management, the construction industry can regain its past glory and reach new heights in a short period of time compared to the old methods.

7.1.5.1 Why is supply chain management important in the construction industry?

The construction industry is extremely unevenly distributed across the world, or it can be said that it is present in patches in any given place. It includes numerous stakeholders, such as direct and indirect suppliers, architects, contractors and so on. It also involves various types of engineers who play specific roles in the completion of projects. To enhance coordination between the different stakeholders involved in a project, supply chain management is required to perform an important role in the construction industry, ensuring that both cost-effectiveness and quality of the project are maintained.

7.1.5.2 How is ethics important in supply chain management?

Ethics are an essential part of supply chain management and due to their importance, supply chain ethics were created. Investors as well as consumers take into consideration the ways any company behaves with its employees, the ways the products are produced by the company and the way the company protects the environment. This leads any company toward the reduction of waste during the production of their products and operations, decreases the negative impact on the environment and increases the positive impact on the environment by using eco-friendly processes and products, in addition to enhancing the working conditions for their employees.

7.1.5.3 Which marketing mix element is related to supply chain management?

The marketing mix element that is related to supply chain management is place because place includes all the processes responsible for providing services as well as the supply of products from the raw materials, which is the initial point to customers, who are the final point.

7.1.6 Objectives

There are various objectives of this chapter, which are as follows:
- This chapter explains supply chain management in general, as well as its application in the construction sector.
- The chapter also highlights the importance of supply chain management in the construction industry.
- It discusses the elements of supply chain management and briefly explains them.
- The chapter further discusses the history and evolution of supply chain management over the decades, explaining the six major movements.
- It briefly describes some major examples of supply chain management in the construction industry.

- The chapter also mentions the various challenges and risks that any organization or company needs to face while implementing supply chain management in construction projects.
- Additionally, this chapter discusses the numerous major benefits that come with implementing construction supply chain management in projects.
- It also discusses the various steps that can be taken to overcome the challenges faced by construction companies or organizations while executing construction supply chain management in their projects.
- The chapter also mentions the major functions of supply chain management in general.

7.2 Motivation

The study was conducted by the author because over several decades, there have been problems in the construction industry. During the recent pandemic, the construction sector was significantly affected in the worst way possible. This further hindered the progress of numerous construction projects in different parts of the world. Additionally, people involved in the construction sector, such as shareholders, had to bear enormous losses. Furthermore, unemployment in the construction sector grew tremendously. The author, being a civil engineer, felt the need to highlight this scenario and the problems of the construction industry. According to the author, the successful implementation of supply chain management in the construction industry will revive the industry quickly, and in a short period of time, the construction sector will reach the same level it had before COVID-19. Moreover, the construction industry can reach new heights with the proper execution of supply chain management with ethics in the future. The author also noticed various problems that occurred in numerous construction projects in the state of Goa in India, where the author resides, as well as in across India. Especially during the recent pandemic, all the progress of ongoing projects stopped, and a lot of mismanagement occurred in many existing construction supply chains. This same scenario also occurred all over the world at different levels simultaneously, particularly during COVID-19. Therefore, it is time to focus on improving the existing supply chain management in the construction industry or creating new ones where they are not present.

7.3 The term's origin, history and evolution over the years

Keith Oliver, a consultant at Booz Allen Hamilton, first introduced the term "supply chain management" in 1982 during an interview with the *Financial Times*. The following year, *WirtschaftsWoche* in Germany published the outcomes of a project led by Wolfgang Partsch called "supply chain management project," which was the first time to see this term applied in a project publicly. By the middle of the last decade of the twentieth century, many books as well as articles were published on the subject "supply chain management" and made it popular among the general public throughout the world. As the time passed, more works have been going on in the direction of supply chain management in the world, and later on, supply chain management software also came into existence to help companies in order to enhance their supply chain performance.

The supply chain management has evolved several times and has developed to its current level due to six main movements, which are:
- Creation era
- Integration era
- Globalization era
- Phase 1 of specialization era
- Phase 2 of specialization era
- Supply chain management 2.0

7.3.1 Creation era

As stated earlier, Keith Oliver was the first person to coin the term "supply chain management" in 1982. Nevertheless, the notion of the supply chain in the field of management was highly significant way back in the first half of the twentieth century, particularly when the assembly line was created. This era was very primitive in nature, and later on, the book published by Handfield and Nichols Jr. in 1999 made this era highly popular among the masses.

7.3.2 Integration era

The development of EDI, or Electronic Data Interchange systems, in the seventh decade of the twentieth century made a significant contribution toward the beginning of the integration era. Later on, when ERP or enterprise resource planning systems were introduced in the last decade of the twentieth century, this era quickly developed. With the growth of Internet-based collaborative systems, the era was supported to continue its development in the twenty-first century.

7.3.3 Globalization era

The era of globalization came into existence when worldwide systems of supplier relationships were given attention, and supply chains expanded beyond the boundaries of a single country to reach other countries and continents. Although many decades ago the oil industry had this feature, the latter half of the ninth decade of the twentieth century saw significant number of organizations that began to incorporate international sources into their core businesses and made this era acknowledged by the world.

7.3.4 Phase 1 of the specialization era

In this era, companies outsourced or subcontracted their distribution in addition to manufacturing systems. In the last decade of the twentieth century, many companies started to concentrate on core competencies as well as specializations. Additionally, they permanently left vertical integration, sold their noncore operations and subcontracted related functions to other companies, which became subcontractors for specific functions. This further altered the needs of management because there was an expansion of the supply chain outside the walls of individual companies in most cases, and there was a distribution of management in which specialized supply chain partners were also included.

7.3.5 Phase 2 of the specialization era

In this era, supply chain management began to be utilized as a service. Although specialization inside supply chain started in the ninth decade but later on supply chain was not only related to logistics and transportation. It also encompassed performance management, partnerships, implementation and even supply planning.

7.3.6 Supply chain management 2.0

Supply chain management 2.0 was built on the previous eras of globalization and specialization. It defines the evolution of methods, processes and tools used for management in the modern era, in addition to changes that occurred within the supply chains. The emergence of TradeCard's supply chain collaboration platform indicated the rise in popularity of collaborative platforms in the modern era. These platforms further helped connect numerous suppliers and buyers with financial institutions, allowing them to utilize automated supply chain financial transactions.

7.4 Literature review

Xie et al. [3] worked on incorporating public-private partnerships (PPPs) and government intervention in environmental transformation. They focused on green supply chain management in the construction industry. They concluded that coordination between the PPP model and government assistance is needed to achieve the desired results.

Sun et al. [4] worked on gray possibility DEMATEL – NK method to find the optimum route to overcome obstacles in implementing green supply chain management in the construction industry. Their results helped them identify the best possible method to execute green supply chain management in the construction sector.

Senthil and Muthukannan [5] worked on developing supply chain risk management in the construction industry with the assistance of an improved neural network. They stated that, based on their study results, there is a need to place strong importance on promotion management as well as standardization.

Lu et al. [6] worked on identifying smart construction objects that can be used as blockchain oracles in supply chain management within the construction sector. They claimed that their study supports and connects blockchain's power in the construction sector.

Hussein and Zyad [7] worked on supply chain management in offsite construction. They stated that methods like offsite construction are both ecological and novel. They studied 309 journal papers on the modeling of offsite construction supply chains to recognize research gaps and trends, which will be helpful for future researchers.

Taghavi et al. [8] worked on detecting and prioritizing the real factors in order to execute green construction supply chain management. They stated that their study results proved that external factors are the most significant and dominant in the green supply chain management process.

Mojumder and Singh [9] conducted an investigative study to adapt green construction supply chain management. They used examples of construction companies in India for their study. They claimed that their study results would assist project managers in the construction sector as well as policymakers focused on clean production.

Battula and Kone [10] conducted a study on the factors directly affecting the execution of construction supply chain management. With the assistance of the RII approach, they were able to analyze the data and provide rankings for the factors based on the RII's mean value.

Xu et al. [11] worked on the construction supply chain management method to minimize the release of acidifying and greenhouse gases from construction materials. They conducted their study using China as a case study. They claimed that their proposed methodology would bring both monetary and environmental benefits after implementation.

Badi and Murtagh [12] worked on green construction supply chain management and systematically presented literature reviews. They also selected 44 papers from the initial 207 papers to analyze them in detail and obtain the desired study results.

Yevu and Darko [13] worked toward digitalizing supply chain management in the construction sector, as well as methods of procurement in the built environment. They stated that their study increases the value and supports digitalization in the construction sector by providing a comprehensive overview of research findings and highlighting the research requirements important for both practitioners and researchers. This advancement is crucial for the development of digital supply chain procurement technologies in a sustainable and digitalized built environment.

Hoseini and Ghannadpour [14] worked on incorporating the scheduling of activities in multiple projects as well as suppliers for green supply chains in the construction sector which are under uncertainty. They used a biobjective linear programming mathematical model to evaluate supply chains in the construction industry and achieve the desired results.

Srikar and Lakshmi [15] worked to develop and evaluate a management model for a sustainable value chain in the construction sector. They used the construction sector in India as a case study. They stated that the data obtained from the project development life cycle research would be helpful for supply chain research as well as design research in the future.

Wei and Sauser [16] studied project management and reviewed supply chain management within the context of the former. They examined numerous published research works that were important for both project management and supply chain management. Additionally, they stated that their study would be helpful in successfully incorporating supply chain management with project management.

Karlsson et al. [17] worked on achieving net-zero carbon emissions in the supply chains of the construction industry. They performed a multidimensional analysis of residential building systems to conduct their study. They claimed that their study results showed that the release of greenhouse gases can be decreased by a maximum of 40% with the assistance of the technologies and practices currently available.

Hamledari and Fischer [18] worked to incorporate financial as well as physical supply chains in the construction sector, where they applied blockchain-based crypto assets to achieve the desired results. They concluded their study after discussing risks such as price fluctuations, safety, regulatory landscapes and possible solutions in their paper.

Bajomo and Zhang [19] worked on a system dynamics method to manage the procurement of construction materials in the construction, engineering and procurement industries. They presented their proposed model of the construction material supply chain in these industries to solve existing problems.

Marzouk and Sabbah [20] worked to find suppliers in the supply chain of the construction industry, using the analytic hierarchy process technique for order preference by similarity to the ideal solution approach to achieve the desired results. They also aimed to propose a computational model of multicriteria decision-making, to introduce it to construction companies for use in the prequalification process to select suppliers.

Su et al. [21] worked on a supply chain level analysis centered on the structural production layer difference method to understand the connection between the release of greenhouse gases, the growth of the construction sector and economic development in China.

Liu et al. [22] worked on incorporating supply chain mitigation options in the construction sector as well as advanced automobile technologies to fulfill the plan of decarbonization. Through their study, they were able to reconsider the release of CO_2 in the road transportation sector and achieve the desired results.

7.5 Main focus of the chapter

This chapter focuses on supply chain management in general as well as in the construction industry. The chapter mentions the functions and elements of supply chain management. Furthermore, it highlights the significance of supply chain management in the construction industry, and with its assistance, construction projects can be completed on time and the quality of the projects can be enhanced. Since the chapter is about supply chain management in the construction industry, it is also a part of supply chains, which is the theme of the book. The chapter also focuses on the numerous major challenges that construction companies face while implementing construction supply chain management in the development of projects as well as various steps to overcome them. Additionally, this chapter highlights the various major benefits that supply chain management brings to the construction sector, which will serve as a key solution to revive the construction industry and help it reach new heights in the future.

7.6 Issues, controversies and problems

There are some major issues with construction supply chain management that need to be taken into consideration. For instance, there are often trust issues between the stakeholders of the construction project. Additionally, designs are frequently changed, which becomes problematic and disrupts ongoing or previous plans. Erroneous calculations on paper create significant problems when implementing them practically in any project. If workers and subcontractors are not adequately trained, there is a high possibility that their inexperience will ruin the project. Poor-quality equipment and construction materials compromise the overall quality of the projects, further reducing their lifespan. Moreover, there are no proper tools to check and measure the performance of all parties involved in the project's development. Additionally, there are no verified or standard construction supply chain models to follow. These problems often create disagreements between the involved parties, such as contractors and subcontractors or contractors and clients. These issues must be resolved to implement successful supply chain management in the construction sector.

7.7 Five major components of supply chain management

The supply chain manager in supply chain management is responsible for coordinating the logistics of all parts of the supply chain, which are as follows:
- The strategy or plan
- The source for either raw materials or services
- The manufacturing of products is concentrated on efficiency as well as productivity
- The logistics part as well as the supply of products to the final consumer
- The system provides return services for undesirable or defective products

The role of a supply chain manager is to reduce costs and shortages as much as possible. Their job is not limited to logistics and purchasing inventory but also includes managing and supervising all operations related to logistics and the supply chain. This is done to make the process highly efficient and reduce the supply chain costs of an organization, according to Salary.com.

In any company, an enhancement in efficiency and productivity can directly impact the lowest level possible. That is why there is a requirement for good supply chain management to keep companies away from potential dangers and chain problems that are directly or indirectly related to them.

7.8 Supply chain management examples in the construction industry

In this section, some major examples of construction supply management are discussed, which are as follows:
- GoFor
- LINKX
- Soil Connect
- Voyage Control
- CEMEX Go

7.8.1 GoFor

GoFor is a major leader in North America and is also part of the CEMEX Ventures investment portfolio. GoFor provides the supply of construction materials to the nearest location of the customer. It is known for providing inexpensive and dependable supplies to its customers, including both large and small orders of construction materials, whether the order is prescheduled or on demand. It has its own platform that assists customers in having a good experience, allowing them to track their orders and associated costs, integrate with their own tools and receive construction materials without the hassle of repeated contact. Founded in 2016 in Canada, it rapidly expanded into major cities in the USA. Currently, it offers its services to numerous industries, adding its busy fleets or selecting it as a logistics partner to provide outsourcing services like supplying construction materials. During the pandemic, the start-up also began supplying masks, hand sanitizers and food to its customers, helping retailers supply their products with the start-up's assistance. This further helped GoFor to enhance its SaaS-based dispatch platform and logistics market, overshadowing others.

7.8.2 LINKX

In the country of Mexico, there is a cloud-based logistics software called LINKX, which supplies products within the country. Its platform optimizes operations with the management of supplies and loads in real time, incorporating all aspects of a given transaction. This start-up allows freight operators to allocate supplies to their own respective carriers or third parties. It also provides advance shipping notices before supplying the materials. After the product is delivered to the designated location, it digitizes the product confirmation and allows for partial delivery confirmation to avoid billing errors or incomplete orders. Additionally, it provides a complete digital report in real time with data-based inferences related to the operations of its customers.

7.8.3 Soil Connect

Soil Connect was one of the top 50 Contech start-ups in 2020. It created a marketplace that links professionals in the construction sector, where one needs to dispose of soil or does not have any use of that soil with the ones who want the soil for their own needs. They developed a highly efficient and superior way of obtaining materials in the construction sector. The platform matches customers based on distance, allowing them to operate over shorter distances, which saves a significant amount of money on transportation, waste treatment and material acquisition. This also considerably decreases the carbon footprint of the construction industry by reducing CO_2 emissions and reutilizing excess materials instead of sending them to landfills. Currently, CEMEX is using this solution in Florida and is trying to expand its use to other markets in the USA. Additionally, it supplies construction materials to its customers for their projects with the help of an app.

7.8.4 Voyage Control

Voyage Control assists developers, construction companies and owners in managing their supply chain and logistics. This includes supplying services to construction sites, end-to-end supply chain management, inventory control, fulfilling compliance needs and supporting payment apps. Its advantages include enhancing driver compliance, restructuring operational processes, helping the environment, minimizing congestion and improving security. By utilizing its platform, businesses can reach beyond sites used solely for construction work. More than 100,000 businesses have utilized its platform to become time-efficient and reduce the wastage of time in their respective processes, reaching more than 4 million supplies throughout Asia-Pacific, North America and Europe. This UK-based start-up was one of the top five winners of the Construction Start-up Competition of 2020.

7.8.5 CEMEX Go

CEMEX Go is the solution and digital platform of CEMEX Ventures, which reorganizes the workflow. This platform allows its customers to manage orders, track their supplies and buy products seamlessly and effortlessly.

CEMEX Ventures is still enthusiastically searching for start-ups and entrepreneurs who can propose solutions and novel business models to optimize the construction supply chain. CEMEX Ventures either invests in these start-ups or partners with them to spread their solutions in different markets. This will ultimately improve supply chain management in the construction industry considerably.

7.9 Digitalization of construction supply chain

Since the construction industry is extremely unevenly distributed, it becomes hard to manage the construction process, particularly in projects where numerous contractors and suppliers are involved. Here, the digital supply chain comes into play to solve this problem by providing assistance to synchronize the supply chain links across several parties involved in the construction project.

The digital technologies used will be able to provide data regarding equipment, construction materials and milestones of the construction project. At the same time, such technologies will be able to coordinate the schedules of numerous parties involved in the project. Furthermore, they can help in collecting details from all the involved parties and allow real-time processes to make decisions based on the collected data.

Then, artificial intelligence as well as Internet of things can be utilized to provide insights into the performance of equipment and assets. These technologies will assist in enhancing quality, minimizing costs and reducing resource consumption in a project.

7.10 Methodology

The basic working of supply chain management in the construction industry is as follows (Figure 7.1): First, there is a need to evaluate the existing supply chain if it is present; second, a new supply chain must be designed; third, the supply chain must be managed properly; and fourth, the supply chain should be constantly enhanced.

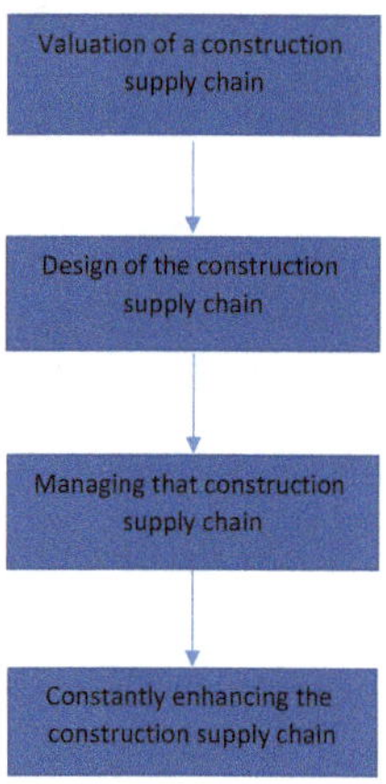

Figure 7.1: The basic working of construction supply chain management [23].

7.11 Results and discussion

Tables 7.1–7.4 and Figures 7.2–7.5 are the collection of data from a number of data sources.

Table 7.1: Comparison between two key performance indicators in order to monitor a supply chain (in %).

Production time	27
Daily performance	40

Source: This is plotted using data acquired and published by Finances Online Reviews for Business [24].

Table 7.2: Comparison between two factors responsible for the increase in supply chain spending in 2018 (in %).

Automation of supply chain management	25
New talent addition	4

Source: This is plotted using data acquired and published by Reuters in 2018, and Finances Online Reviews for Business [24].

Table 7.3: Supply chain changes faced by professionals in 2020 (in %).

Retailers were forced to adjust their respective supply chains for e-commerce	64
The pandemic heavily disrupted the supply chains of retailers globally	12

Source: This is plotted based on data acquired and published by RetailNext in 2020, and Finances Online Reviews for Business [24].

Table 7.4: Comparison between two causes that disrupt supply chain in the USA (in %).

Loss of talent	51
IT outages that are not planned	68

Source: This is plotted using data acquired and published by BCI in 2020, and Finances Online Reviews for Business [24].

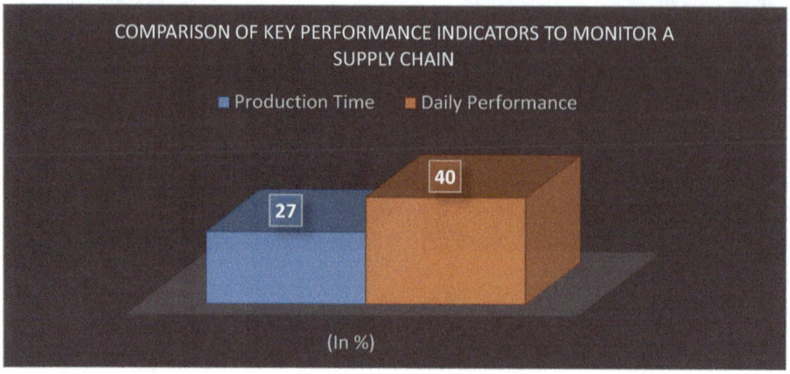

Figure 7.2: Comparison of two key performance indicators in order to monitor a supply chain.
Source: This graph is plotted using data acquired and published by Finances Online Reviews for Business [24].

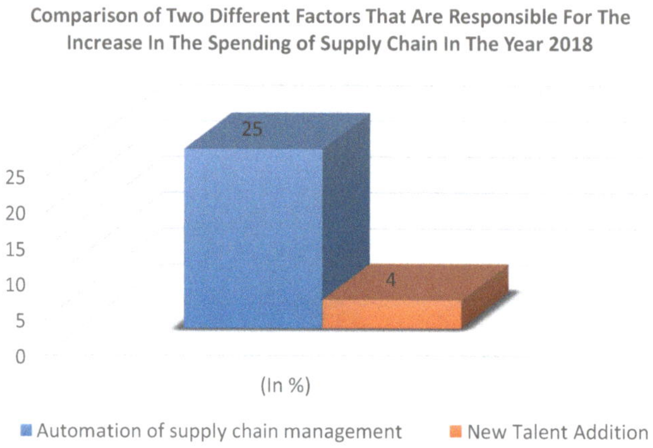

Figure 7.3: Comparison between two factors that are responsible for the increase in the spending of supply chain in 2018.
Source: This graph is plotted using data acquired and published by Reuters in 2018, and Finances Online Reviews for Business [24].

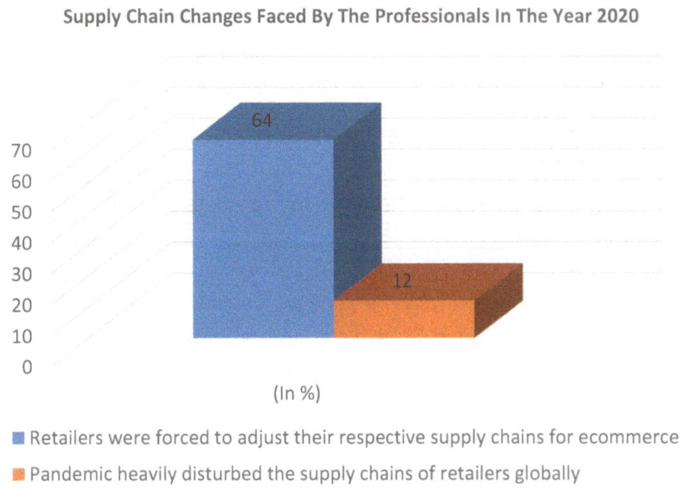

Figure 7.4: Supply chain changes faced by professionals in 2020.
Source: This graph is plotted based on data acquired and published by RetailNext in 2020, and Finances Online Reviews for Business [24].

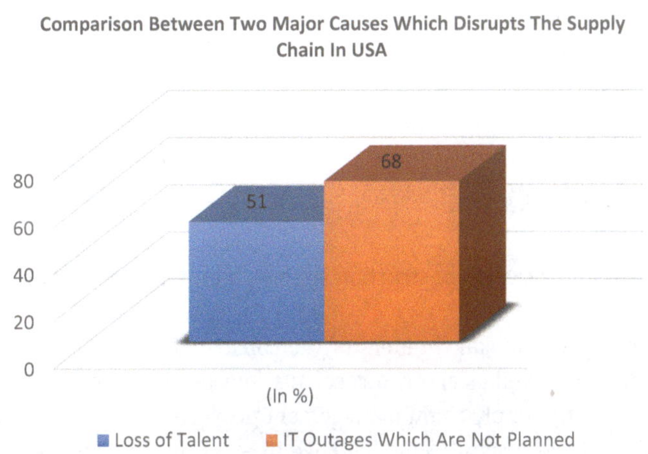

Figure 7.5: Comparison between two causes that disrupts supply chain in the USA.
Source: This graph is plotted using data acquired and published by BCI in 2020, and Finances Online Reviews for Business [24].

7.12 Advantages of construction supply chain management

There are several major advantages of using supply chain management in the construction industry, which are as follows:

1. Supply chain management ensures that the timelines of a project are set in a realistic and efficient manner, and that targets are reached within the given timeframe. Consequently, it enhances the satisfaction level of the concerned clients as well as the reputation of the organization in the market.
2. A competitive advantage is provided by the supply chain management to organizations by giving them the ability to manage numerous variables and factors in a given project in a highly efficient and effective manner compared to their competitors.
3. If supply chain management is properly integrated with ethics by all the stakeholders of a project, then their mutual trust can increase considerably, creating a high possibility of them working together on future construction projects.
4. The uncertainty associated with logistics, equipment and materials can be minimized by organizations or companies with the assistance of supply chain management. This can be achieved by creating long-term alliances within the supply chain. This approach will minimize costs in both current and future projects, as the involved parties will prioritize mutual benefits and trust in the long run over personal gains in the short run.

7.13 Challenges and risks

In this section, the challenges and risks in construction supply chain management are discussed, as follows:

1. The absence of trust between the stakeholders of the construction project, frequent changes in the design as well as erroneous calculations can become significant obstacles in executing the supply chain management properly.
2. The construction industry faces a significant challenge in executing supply chain management successfully due to the absence of any standard or verified supply chain management model.
3. Insufficient training of workers and subcontractors as well as the low quality of equipment and construction materials are also major challenges in implementing supply chain management in the construction sector.
4. Then, a common challenge is the lack of cooperation and partnership among all the involved parties in the construction project.
5. There are no proper tools that can effectively measure the performance of all parties involved in the construction project.

7.14 Ways to overcome the challenges in the implementation of construction supply chain management

There are several major steps that can be taken to overcome the challenges of supply chain management in the construction sector, which are as follows:

1. Heightening trust and increasing the communication between stakeholders of the project are important in managing unforeseen events and risks that occur throughout the development of the construction project. This is essential for implementing supply chain management in the construction industry.
2. When all the involved parties are amenable to alterations across all parts of the supply chain, only then construction supply chain management in the construction industry can be successfully executed.
3. Subcontractors must also be efficiently managed as they should be considered a part of the supply chain. Additionally, input from subcontractors must be sought in order to increase the project's value.
4. With the help of a cultural shift and by ensuring that all involved parties understand how supply chain management can generate profits and bring various benefits to the project, the successful implementation of supply chain management is possible in the construction industry.
5. There is a need to develop specific parameters to use as a standard for forming alliances with supply chain partners. At the same time, it is essential to monitor their performance in the project, which will aid in the proper implementation of construction supply chain management.

7.15 Limitations

This chapter focuses on supply chain management in general as well as in the construction sector. It briefly highlights the functions of supply chain management, its history and its evolution over the years. It also mentions the importance of supply chain management in the construction industry, its basic workings and its benefits. However, the chapter does not cover the situation of supply chain management in the construction industry in any particular country. There are many examples of construction supply chain management worldwide, but in this chapter, only five major examples are briefly covered.

7.16 Solutions and recommendations

The supply chain management in the construction industry brings numerous benefits but it also faces several challenges that cannot be ignored. The author has mentioned the steps to overcome these challenges with the available technologies. However, with more research in this area and the invention and development of new technologies, construction supply chain management will improve further, and there will be more innovative ways to overcome all possible challenges in the future. The author recommends that all construction companies and involved parties in all construction projects understand the significance of construction supply chain management and properly implement it in the development of their respective projects.

7.17 Future research directions

Supply chain management in the construction industry is important to revive the industry after the recent pandemic. Additionally, this will be a major step toward helping the construction industry reach new heights in the future. Hence, more research is required in the field of construction supply chain management to improve it further and enhance the condition of the construction supply chain as well as the construction industry. Future research in this area will help contractors, suppliers, engineers, subcontractors, architects, clients, etc., while working to develop construction projects and their quality in the future.

7.18 Conclusion

This chapter explained supply chain management in general terms as well as construction supply chain management briefly. It also highlighted the functions of supply chain management and supply chain resilience. It then discussed the history and evolution of supply chain management and also explained the elements of supply chain management. The chapter further explained the basic working of construction supply chain management and the benefits it can provide to the construction industry. Additionally, some major challenges and risks are mentioned in the chapter that construction companies face while implementing supply chain management in the development of construction projects. Furthermore, the author discussed some major steps that will help overcome the existing challenges of construction supply chain management.

References

[1] Lambert, D. M., & Cooper, M. C. Issues in supply chain management. Industrial marketing management, 29(1), 65-83, 2000.

[2] Bowersox, Donald J., & Closs, David J. Logistical Management: The Integrated Supply Chain Process. McGraw-Hill College. ISBN-13: 978-0070068834

[3] Z. Xie, Chen and Allen. Green construction supply chain management: Integrating governmental intervention and public–private partnerships through ecological modernisation. Journal of Cleaner Production, 331:129986, January 2022.

[4] M. Sun, Dang and Xu. Optimum path for overcoming barriers of green construction supply chain management: A grey possibility DEMATEL-NK approach. Computers & Industrial Engineering, 164:107833, February 2022.

[5] Senthil and Muthukannan. Development of lean construction supply chain risk management based on enhanced neural network. Materials Today: Proceedings, 56, Part 4, 2022, Pages 1752–1757.

[6] L. Lu, Z. Xue, Wu and Yeh. Exploring smart construction objects as blockchain oracles in construction supply chain management. Automation in Construction, 129:103816, September 2021.

[7] Hussein and Zayed. Modelling in off-site construction supply chain management: A review and future directions for sustainable modular integrated construction. Journal of Cleaner Production, 310:127503, 10 August 2021.

[8] F. Taghavi, Wong and Hoseini. Identifying and prioritizing the effective factors in the implementation of green supply chain management in the construction industry. Sustainable Operations and Computers, 2:97–106, 2021.

[9] Mojumder and Singh. An exploratory study of the adaptation of green supply chain management in construction industry: The case of Indian Construction Companies. Journal of Cleaner Production, 295, 126400, 1 May 2021.

[10] N. Battula and Kone. A study on factors involved in implementation of supply chain management in construction industry. Materials Today: Proceedings, 33(Part 1):446–449, 2020.

[11] X. Huang, Shi and Deng. Supply chain management approach for greenhouse and acidifying gases emission reduction towards construction materials industry: A case study from China. Journal of Cleaner Production, 258:120521, 10 June 2020.

[12] Badi and Murtagh. Green supply chain management in construction: A systematic literature review and future research agenda. Journal of Cleaner Production, 223, 312–322, 20 June 2019.

[13] Y. Yevu and Darko. Digitalization of construction supply chain and procurement in the built environment: Emerging technologies and opportunities for sustainable processes. Journal of Cleaner Production, 322:129093, 1 November 2021.

[14] N. Hoseini and Ghannadpour. Integrated scheduling of suppliers and multi-project activities for green construction supply chains under uncertainty. Automation in Construction, 122:103485, February 2021.

[15] R. Srikar and Lakshmi. Evaluation and development of management model for sustainable value chain in Indian construction industry. Materials Today: Proceedings, December 2021.

[16] P. Wei and Sauser, Review of supply chain management within project management. Project Leadership and Society, 2:100013, December 2021.

[17] R. Karlsson, Johnsson and Erlandsson, Achieving net-zero carbon emissions in construction supply chains – A multidimensional analysis of residential building systems. Developments in the Built Environment, 8, September 2021, 100059.

[18] Hamledari and Fischer. The application of blockchain-based crypto assets for integrating the physical and financial supply chains in the construction & engineering industry. Automation in Construction, 127:103711, July 2021.

[19] O. Bajomo and Zhang. A systems dynamics approach to the management of material procurement for engineering, procurement and construction industry. International Journal of Production Economics, 244:108390, February 2022.

[20] Marzouk and Sabbah. AHP-TOPSIS social sustainability approach for selecting supplier in construction supply chain. Cleaner Environmental Systems, 2:100034, June 2021.

[21] Z. Su, Ma and Ji. Understanding the relationships between the development of the construction sector, carbon emissions, and economic growth in China: Supply-chain level analysis based on the structural production layer difference approach. Sustainable Production and Consumption, 29:730–743, January 2022.

[22] W. Liu, H. Lyu, Yang and Gao. Rethinking the carbon dioxide emissions of road sector: Integrating advanced vehicle technologies and construction supply chains mitigation options under decarbonization plans. Journal of Cleaner Production, 321:128769, 25 October 2021.

[23] http://www.aasmr.org/jsms/Vol8/Vol.8_No.4_1.pdf

[24] 97 Supply Chain Statistics You Must Know: 2021/2022 Market Share Analysis & Data, https://financesonline.com/supply-chain-statistics/

Vilda Purutçuoğlu* and Başak Bahçivancı

Chapter 8
Selection of threshold in binary graphs of biological networks

Abstract: In recent years, accurately screening genes and their interactions has become increasingly important for personalized medicine. Understanding and detecting gene interactions are pivotal, but discovering these interactions is challenging due to the inherent structural and functional complexities of biological systems. These challenges include the high sparsity of gene interaction networks, the disparity between the number of genes and the number of samples, and the strong correlations among genes. The Gaussian graphical model (GGM) is a fundamental tool used to infer gene interaction networks. It represents relationships between biological entities through an undirected graph, where nodes denote genes and edges indicate conditional dependencies. To construct these graphs, the precision matrix, which contains conditional dependencies, is converted into a binary form, with 0 representing the absence of an edge and 1 indicating the presence of an edge. This conversion requires thresholding, a critical step that affects network construction and subsequent developments. Given the importance of network-based biological research for discovering novel biomarkers and drugs, threshold selection for the precision matrix has become a focal point. Various methods for selecting the optimal threshold are discussed in the literature, including both parametric and nonparametric approaches. For instance, Schneider et al. [4] propose a parametric threshold selection based on data distribution and a nonparametric method using Hill's estimator for univariate extreme value analysis.

Keywords: Gene Screening, Gene Interactions, Personalized Medicine, Gaussian Graphical Model (GGM), Precision Matrix

Acknowledgments: The authors would like to thank Meral Ebegil, Erdem Karabulut, Ömür Uğur and Barış Sürücü in the interpretation of the analyses and the Middle East Technical University research grant (no. BAP-08-11-2017-035) for their support.

***Corresponding author: Vilda Purutçuoğlu,** Department of Statistics, Middle East Technical University, 06800 Ankara, Turkey, e-mail: vpurutcu@metu.edu.tr
Başak Bahçivancı, Earlham Institute, Norwich Research Park, Norwich NR47UZ, UK, e-mail: basakbahcivanci@gmail.com

https://doi.org/10.1515/9783110717655-008

8.1 Introduction

In recent years, accurately screening genes and their interactions have become more crucial since understanding and precisely detecting gene interactions hold the key to personalized medicine. However, the discovery of gene interactions faces several challenges due to the structural and functional complexities inherent in biological systems, such as the high sparsity of the networks, the high number of genes relative to the number of samples collected and the high correlation between genes. One of the fundamental models to infer these networks is the Gaussian graphical model (GGM), which is used to explain the relationship between two biological entities under the steady-state activation of the system via an undirected graph [1]. By using GGM, a graphical representation can be built, where edges indicate interactions and nodes represent genes. Moreover, these edges indicate the conditional dependencies between those genes. To build such graphs, precision matrix containing conditional dependencies is converted into a binary form, where 0 represents the absence of an edge and 1 denotes the presence of an edge. To be able to convert these numerical values into their binary forms, thresholding is needed. Since biological researches, which are network-based, show the cruciality to develop novel biomarkers and drugs [2], threshold selection for precision matrix has gained more attention due to its direct influence on the construction of networks and developments which follow this step.

Consequently, to obtain more reliable results regarding the activations of complex systems, several methods for selecting the optimal threshold value are suggested in the literature [3]. Some of these methods are based on parametric approaches, while others are fully nonparametric. For example, Schneider et al. [4] select the threshold parametrically by considering the distribution of the data. They also adopt an optimization approach based on an estimator called Hill to propose a nonparametric threshold selection method in univariate extreme value analysis. On the other hand, a nonparametric procedure that depends on selecting the optimal p-value for the edges' significance via hypothesis testing is done by comparing existing edges via their randomness. This approach is suggested by Almıdevar [5]. Moreover, Chen et al. [6] propose a parametric method that utilizes a mixture of distributions in the exponential family. There are also other studies that implement empirical analyses for the underlying system [7] or intuitively select 0.5 or 0.6 as a constant term to control biological network sparsity [8].

Considering all the methods mentioned above, it is evident that parametric approaches require detailed information about the data and have restrictions regarding the underlying distribution. In contrast, nonparametric approaches are flexible. Nonetheless, they can be computationally demanding if their calculations are based on optimization techniques. However, methods that depend on a fixed value or accuracy measures to validate the estimated systems' structure via their (quasi) representations are known to be computationally user-friendly. On the other hand, current methods in this group do not consider the topological features of the network of interest.

Hence, in this study, we initially assess the performance of major threshold selection methods in the literature, which are typically applied for binary construction. Then, we present an alternative approach and evaluate the accuracies of the estimates under distinct measures and computational time. In this evaluation, we use GGM as it is the fundamental model of many complex modeling approaches.

8.2 Methods

8.2.1 Gaussian graphical models

One of the efficient ways to represent interactions between two entities of a biological system is through graphical models. In these models, there is a set of nodes that represent biological entities such as proteins or genes, and edges between those biological entities, which are called interactions. A system that consists of p nodes is represented by a random variable, which can be written as $Y = (Y^{(1)}, Y^{(2)}, \ldots, Y^{(p)})$. In GGM, it is assumed that Y has multivariate normal distribution with mean vector μ and covariance-variance matrix Σ. Here, the inverse of Σ has a special name called precision matrix, and is denoted by θ and it is important since it gives conditional independencies θ_{ij} of the corresponding two nodes. Accordingly, θ_{ij}, which is greater than the selected threshold, is denoted by 1, and θ_{ij}, which is less than the selected threshold, is shown by 0. If an entry of θ_{ij} is 0, then the variable corresponding to the ith row and the jth column is conditionally independent given the other variables. There are several methods to estimate the components of the GGM. One of these approaches is based on the maximum likelihood estimator (MLE). This well-known approach cannot be directly applicable in graphical representation of the biological networks because the structure of θ is very sparse and its entries/components, i.e., θ_{ij}'s $(i, j = 1, \ldots, p)$, cannot be inferred directly. The graphical lasso (GLASSO) is the adopted version of the MLE, where L_1-norm penalty is used to control θ_{ij}. For more details about the computation steps of θ_{ij}, we can recommend Whittaker [1] and Friedman et al. [9]. In this study, we perform GLASSO for the inference of θ due to its common usage and simplicity. As our proposal threshold value selection is free from the estimation approach of GGM coefficients, it can be used for other inference techniques as well, such as neighborhood selection, birth-death approach or its Bayesian alternatives [10–12]. Hence, in the R programming, the glasso method by the huge package is applied to construct the graph. On the other hand, due to the optimization problem that is originated from the sparsity of biological networks, there is a necessity to select the best-fitted GGM among estimated ones. There are several model selection criteria supported by huge package such as the Bayesian information criterion, the Akaike information criterion, the rotation information criterion and the stability approach to regularization selection (StARS). Among these alternatives, we choose the StARS [13] criterion since it is specifically designed for sparse models.

8.2.2 Threshold selection methods

Once coefficients θ are estimated, they are converted to binary values since the binary values are used to generate an adjacency matrix A, which indicates the presence or absence of interactions between network elements via 1 and 0 values, respectively. As previously mentioned, there are many approaches to convert the values of θ_{ij} in θ into 1 and 0 values in such a way that θ_{ij}, which is greater than the selected threshold, is denoted by 1, and θ_{ij}, which is less than the selected threshold, is shown by 0. In this study, we use the most common threshold selection methods, such as the kappa maximized threshold (KMT) criterion, maximized sum threshold (MST), minimized difference threshold (MDT) and 0.5 T value criterion, to compare the performance of our proposed method with these alternative approaches. These methods are well-known and widely used due to their high accuracy or their convenience in application. In addition to these methods, we propose a novel nonparametric threshold selection criterion that considers the topology of the network system.

8.2.3 Kappa maximized threshold criteria

KMT criterion is an approach which maximizes the kappa statistics [14]. Kappa statistic is calculated via

$$\kappa = \frac{p_0 - p_e}{1 - e} \tag{8.1}$$

where p_0 stands for the observed percentage of the agreement (accuracy) and p_e denotes the expected agreement probability due to chance. These probabilities are calculated by

$$p_0 = \frac{\mathrm{TP} + \mathrm{TN}}{n}$$

and

$$p_e = \frac{(\mathrm{TP} + \mathrm{FN}) \times (\mathrm{TP} + \mathrm{FP}) + (\mathrm{TN} + \mathrm{FN}) \times (\mathrm{TN} + \mathrm{FP})}{\mathrm{n}^2}.$$

Here, TP (true positive) is the number of agreements in the entities of both the true adjacency matrix and the estimated adjacency matrix on the existence of an edge, i.e., 1. TN (true negative) denotes the number of agreements on the absence of an edge, i.e., 0, in the entities of both the true adjacency matrix and the estimated adjacency matrix. FP (false positive) represents the number of incorrect predictions of the actual absence of an edge, and FN (false negative) shows the number of incorrect predictions of the actual presence of an edge. A simple representation of these values is given in Table 8.1. The method calculates a kappa value for each 100-threshold value, and the one that provides the maximum kappa statistic is selected as the threshold [14–16]. In this study, the 100 candidate threshold values, which are between the minimum and the maximum entries of the precision matrix, are calculated via an increment whose calculation is done by the following formula:

$$\text{increment} = \frac{\max\left(\text{entries of } S^{-1}\right) - \min\left(\text{entries of } S^{-1}\right)}{100} \tag{8.2}$$

where S is the sample covariance-variance matrix.

Table 8.1: The general confusion matrix.

Predicted class	**Actual class**	
	Positive	**Negative**
Positive	TP	FP
Negative	FN	TN

8.2.4 Maximized sum threshold

The MST [17] is based on the calculation of the sum of specificity and sensitivity, which are computed by

$$\text{Specificity} = \frac{\text{TN}}{\text{TN} + \text{FP}} \tag{8.3}$$

$$\text{Sensitivity} = \frac{\text{TP}}{\text{TP} + \text{FN}} \tag{8.4}$$

In these expressions, specificity indicates the proportion of actually nonexisting edges that are correctly identified, whereas sensitivity computes the proportion of actually existing edges that are correctly identified.

After calculating the sum of specificity and sensitivity for each 100-candidate threshold value, which is calculated in the same manner as in the KMT criterion, MST selects the one which maximizes the sum as the optimal threshold [16].

8.2.5 Minimized difference threshold

The MDT criterion [16] calculates the difference between the specificity and the sensitivity for each 100-candidate threshold, which is calculated in a similar manner as in the previous two methods, i.e., KMT criterion and MST. Accordingly, this method selects the value that gives the minimum difference between the specificity and the sensitivity (considering their absolute values) as its optimal threshold value [16].

8.2.6 0.5 T criterion

This method is considered a subjective approach in the literature since it neither uses a specific index, such as kappa, nor implements a trade-off between two properties, such as sensitivity and specificity [18]. Instead, it simply selects the optimal threshold as 0.5 for all conditions without considering the structure of networks or the performance of accuracy measures.

8.2.7 Proposed threshold selection method

Rather than using a specific index or a trade-off between two properties, we propose an easy threshold selection method that depends on the topology of the network. For this purpose, we use information about the sparsity level of networks, which can vary for each network of interest. Hence, the steps of our procedure can be listed as follows:

Step 1: Calculate the sparsity percentages that are specific to the network classes.
Step 2: Construct the GGM model via GLASSO and obtain the precision matrix θ.
Step 3: Sort the entries of the precision matrix from the smallest to the largest.
Step 4: Find the threshold value by imposing the sparsity level on the vector composed of the sorted entries of the precision matrix.
Step 5: Construct the adjacency matrix by imposing the threshold value on the precision matrix.
Step 6: Obtain the accuracy measures by using the adjacency matrix found in the previous step.

For Step 1, the sparsity percentages of the random and scale-free networks' means are calculated under 300 Monte Carlo runs by generating random and scale-free networks using the huge R package. These two networks are the main topological structures of biological networks [19, 20]. On the other hand, from our preliminary analyses, it is observed that the dimension of the network is another characteristic feature that can affect the sparsity percentage.

Therefore, we take this effect into account in our analyses. For our assessments using simulated data, we consider the dimensions of the system to be 20, 50 and 100 for both random and scale-free structures, assuming that the networks have small (i.e., 20), medium (i.e., 50) and large (i.e., 100) dimensions. We calculate the mean sparsity percentage of the random network under 300 Monte Carlo runs for system dimensions of 20, 50 and 100 as 0.854, 0.942 and 0.97, respectively. We also compute the same values for the scale-free network under 300 Monte Carlo runs for system dimensions of 20, 50 and 100 as 0.905, 0.961 and 0.98, respectively. Then, we estimate our model parameters, i.e., the entries of θ in Step 2.

In Step 3, we sort the entries of Θ, from smallest to largest by creating a vector. Then, in Step 4, by multiplying the length of the vector with the sparsity percentage, we get the vector index corresponding to the cut-off point that will satisfy the sparsity percentage of the underlying network. In this way, we can get the element of the vector that denotes the vector index. The element that we get from the vector is simply one of the entries of the precision matrix, and it also corresponds to our optimal threshold value that satisfies the sparsity percentage of the underlying network structure imposed on the precision matrix.

Here, in Step 5, by using the threshold value that optimizes the precision matrix' sparsity level, we convert the precision matrix into the adjacency matrix in the same manner as applied in the literature. That is, the entries that exceed the threshold value are converted into 1, indicating an interaction between two nodes, and the entries below the threshold value are converted into 0, representing no interaction between two nodes. In Step 6, the accuracy measures are calculated using this generated adjacency matrix.

8.3 Analysis via simulated datasets

In our analyses, we generate six datasets from a multivariate normal distribution under the dimensions of the system as 20, 50 and 100 number of genes for both random and scale-free structures. Moreover, we simulate 50 observations for each gene. Then, based on 1,000 Monte Carlo runs, we compare the mean accuracy value, F-measure and Matthew's correlation coefficient (MCC), whose expressions are denoted below. In this calculation, when MCC is closer to 1, it is interpreted as high accuracy. By contrast, when it is closer to −1, it is interpreted as low accuracy. For other measures, 1 indicates perfection and 0 implies the worst performance:

$$\text{Accuracy} = \frac{\text{TP} + \text{TN}}{\text{TP} + \text{TN} + \text{FP} + \text{FN}} \tag{8.5}$$

$$\mathrm{MCC} = \frac{(\mathrm{TP} \times \mathrm{TN}) - (\mathrm{FP} \times \mathrm{FN})}{\sqrt{(\mathrm{TP} + \mathrm{FP})(\mathrm{TP} + \mathrm{FN})(\mathrm{TN} + \mathrm{FP})(\mathrm{TN} + \mathrm{FN})}} \tag{8.6}$$

$$F\text{-measure} = 2\frac{\mathrm{Precision} \times \mathrm{Recall}}{\mathrm{Precision} + \mathrm{Recall}} \tag{8.7}$$

where

$$\mathrm{Precision} = \frac{\mathrm{TP}}{\mathrm{TP} + \mathrm{FP}} \text{ and } \mathrm{Recall} = \frac{\mathrm{TP}}{\mathrm{TP} + \mathrm{FN}} \tag{8.8}$$

Accordingly, the outcomes of random and scale-free scenarios are given in Tables 8.2, 8.4 and 8.6, and Tables 8.3, 8.5 and 8.7, respectively. From the tables, it is seen that the MST and KMT methods have the best accuracy for both network types. However, KMT method has better computational time in terms of CPU (central processing unit) than MST method. Furthermore, the proposed method is the best in terms of *F*-measure and MCC, and it has significantly better computational time compared to KMT method under low dimensional systems. Additionally, 0.5 T criterion has the worst MCC, and it cannot calculate *F*-measure since it fails to detect any TP (any interaction). This result is interesting because of the fact that 0.5 T criterion is the most common selection approach in the analyses [16].

Table 8.2: GGM via graphical lasso results based on 1,000 Monte Carlo runs for random networks having 20 genes.

Methods	Accuracy	*F*-measure	MCC	User CPU time
Proposed method	0.837	0.312	0.215	0.065
MST	0.839	0.289	0.212	5.818
MDT	0.835	0.283	0.188	5.834
KMT	0.839	0.289	0.212	5.663
0.5 T criterion	0.809	–	−0.093	0.062

Table 8.3: GGM via graphical lasso results based on 1,000 Monte Carlo runs for scale-free networks having 20 genes.

Methods	Accuracy	*F*-measure	MCC	User CPU time
Proposed method	0.878	0.324	0.248	0.066
MST	0.881	0.320	0.256	5.709
MDT	0.878	0.315	0.240	5.741
KMT	0.881	0.320	0.256	5.524
0.5 T criterion	0.855	–	−0.074	0.060

Moreover, from the results in Tables 8.4 and 8.5, it is evident that the proposed method is slightly better in terms of detecting correctly classified interactions, which is presented as the accuracy value in the tables. On the other hand, the 0.5 T criterion remains the worst for predicting interactions with its negative MCC and lowest accuracy value compared to other methods. Furthermore, the *F*-measure of 0.5 T is not applicable since the method is unable to identify any correctly classified interactions, represented as TP. On the other hand, from the comparison of computational demand, it is observed that the proposed approach has the second-best efficiency after the 0.5 T criterion, and the difference in CPU usage between the 0.5 T criterion and the proposed approach is very small even though other methods require significantly higher computational demand.

Overall, the results in Tables 8.4 and 8.5 show that the proposed method can be a reasonable threshold criterion in terms of accuracy and computational demand compared to other criteria under both network types in medium dimensions.

Table 8.4: GGM via graphical lasso results based on 1,000 Monte Carlo runs for random networks having 50 genes.

Methods	Accuracy	*F*-measure	MCC	User CPU time
Proposed method	0.932	0.271	0.247	0.538
MST	0.930	0.312	0.280	34.271
MDT	0.930	0.312	0.280	34.063
KMT	0.930	0.312	0.280	35.408
0.5 T criterion	0.921	–	−0.036	0.342

Table 8.5: GGM via graphical lasso results based on 1,000 Monte Carlo runs for scale-free networks having 50 genes.

Methods	Accuracy	*F*-measure	MCC	User CPU time
Proposed method	0.954	0.149	0.180	0.538
MST	0.943	0.249	0.220	35.144
MDT	0.942	0.245	0.217	33.837
KMT	0.943	0.251	0.223	33.725
0.5 T criterion	0.941	–	−0.029	0.347

Finally, Tables 8.6 and 8.7 present the Monte Carlo runs for the random and scale-free networks under high-dimensional systems. From the tabulated values, it is observed that the MST, MDT and KMT criteria have the best accuracy; however, our proposed method has a better *F*-measure, MCC and significantly much better CPU time for random network. On the other hand, in Table 8.7, it is seen that KMT is the best in terms of *F*-measure and MCC, whereas it has worse CPU time compared to our proposed method and 0.5 T criterion for scale-free network. Furthermore, it is clear that although 0.5 T criterion has the best accuracy and CPU time, it dramatically fails to catch the network structure since it cannot predict any interaction at all and has the worst MCC. Overall, the proposed method has high accuracy measures as KMT and less computational time as 0.5 T criterion for high dimensional systems too.

Table 8.6: GGM via graphical lasso results based on 1,000 Monte Carlo runs for random networks having 100 genes.

Methods	Accuracy	*F*-measure	MCC	User CPU time
Proposed method	0.958	0.292	0.273	1.609
MST	0.960	0.273	0.255	154.009
MDT	0.960	0.273	0.255	154.066
KMT	0.960	0.273	0.255	153.868
0.5 T criterion	0.960	–	−0.018	1.605

Table 8.7: GGM via graphical lasso results based on 1,000 Monte Carlo runs for scale-free networks having 100 genes.

Methods	Accuracy	*F*-measure	MCC	User CPU time
Proposed method	0.967	0.166	0.149	1.452
MST	0.965	0.174	0.158	141.767
MDT	0.965	0.173	0.157	141.798
KMT	0.966	0.175	0.159	141.746
0.5 T criterion	0.970	–	−0.014	1.449

8.4 Discussion

In this study, we have evaluated the threshold selection methods using simulated datasets under distinct dimensions and used GGM, whose precision matrix is estimated by the GLASSO approach. The construction of adjacency matrix that is generated from the estimated precision matrix of the GGM has some ad hoc calculations because the selection of the threshold value, which converts the precision matrix to the adjacency matrix, is mainly based on empirical results that can validate the biological knowledge about the systems. It can be also assigned as a fixed value such as 0.5 due to its simplicity [16], whereas such selection can have an undeniable impact on accuracy of the model and can give nonsense results with unbalanced samples [18–21].

Hence, we have proposed an alternative threshold selection method. Our method considers the topology of the underlying networks and suggests an objective way for threshold selection. From simulation studies, it has been seen that our proposed method can efficiently detect the network structure when compared to other well-known methods for both random and scale-free networks. In addition, the proposed method and the 0.5 T criterion have the lowest CPU time when compared to other methods. Indeed, even though the 0.5 T criterion is the most common approach in the literature, it has the worst prediction performance while detecting the interactions of the network for both network classes. Therefore, we believe that our proposed method can be a promising approach in the field due to its accuracy, computational efficiency and simplicity in calculation. Finally, we believe that this idea can also be adapted to other network models that need to convert the estimated strengths of the edges to binary values.

References

[1] J. Whittakeri. Graphical Models in Applied Multivariate Statistics. Wiley, 1991.

[2] D. H. Roukos. Networks medicine: From reductionism to evidence of complex dynamic biomolecular interactions. Pharmacogenomics, 12(69):5–8, 2011.

[3] D. S. Bassett and E. Bullmore. Small-world brain networks. Neuroscientist, 12(5):12–23, 2006.

[4] L. F. Schneider, A. Krajina and T. Krivobokova. Threshold selection in univariate extreme value analysis. Extreme, 24:881–913, 2021.

[5] A. Almıdevar. Selection of statistical thresholds in graphical model. EURASIP Journal of Bioinformatic and Systems Biology, 878013, 2010.

[6] S. Chen, D. M. Witten and A. Shojaie. Avoiding negative populations in explicit Poisson tau-leaping. Biometrika, 102:47–64, 2015.

[7] F. Liu, S. W. Zhang, W. F. Guo, F. G. Wei and L. Chen. Inference of gene regulatory network based on local Bayesian networks. PLOS Computational Biology, e1005024, 2016.

[8] S. M. Gibson, S. P. Ficklin, S. Isaacson, F. Luo, F. A. Feltus and M. C. Smith. Massive-scale gene co-expression network construction and robustness testing using random matrix theory. PLOS One, 80(e55871), 2013.

[9] J. Friedman, T. Hastie and R. Tibshirani. Sparse inverse covariance estimation with the graphical lasso. Biostatistics, 9:432–441, 2008.

[10] A. Mohammadi and E. C. Wit. Bayesian structure learning in sparse Gaussian graphical models. Bayesian Analysis, 10, 2015.

[11] H. Farnoudkia and V. P. Uǒglu. Copula Gaussian graphical modelling of biological networks and Bayesian inference of model parameters. Scientia Iranica, 26(2):495–505, 2019.

[12] A. Dobra and A. Lenkoski. Copula Gaussian graphical models and their application to modelling functional disability data. The Annals of Applied Statistics, 5(9):69–93, 2011.

[13] H. Liu, K. Roeder and L. Wasserman. Stability approach to regularization selection (stars) for high dimensional graphical models. Advances in Neural Information Processing Systems, 24(2):1432–1440, 2010.

[14] A. Guisan, J. P. Theurillat and F. Kienast. Predicting the potential distribution of plant species in an alpine environment. Journal of Vegetation Science, 9:65–74, 1998.

[15] W. Thuiller, S. Lavorel, M. B. Araujo, M. T. Sykes and I. C. Prentice. Binomial leap methods for simulating stochastic chemical kinetics. Proceedings of the National Academy of Science, 102(82):45–50, 2005.

[16] A. Jiménez-Valverde and J. M. Lobo. Threshold criteria for conversion of probability of species presence to either–or presence–absence. Acta Oecologic, 31(36):1–9, 2007.

[17] S. Manel, H. C. Williams and S. J. Ormerod. Evaluating presence–absence models in ecology: The need to account for prevalence. Ecography, 28(3):21–31, 2005.

[18] C. Liu, P. M. Berry, T. P. Dawson and R. G. Pearson. Pearson, selecting thresholds of occurrence in the prediction of species distributions. Acta Oecologic, 31(36):1–9, 2005.

[19] A. L. Barabasi. Network Science. Cambridge University Press, 2016.

[20] A. L. Barabasi and Z. N. Oltvai. Network biology: Understanding the cell's functional organization. Nature Review of Genetic, 5(10):1–13, 2004.

[21] J. S. Cramer. Logit Models from Economics and Other Fields. Cambridge University Press, 2003.

Mehmet Ali Kaygusuz and Vilda Purutçuoğlu

Chapter 9 Model selection criteria with bootstrap algorithms: applications in biological networks

Abstract: Model selection methods play a fundamental role in both classical and modern statistical theories. Owing to the technological advancements, we now have more data in fields such as engineering, medicine and finance. Consequently, the estimation in high-dimensional graphical models and the selection of the regularization parameter embedded in these models under high dimensions have become more important. Therefore, many model selection criteria have been suggested in the literature such as "Akaike's information criterion" and the "Bayesian information criterion" as well-known approaches, as well as some extended versions, such as the "consistent Akaike information criterion with Fisher information" and "information and complexity selection" methods. In this study, we used the aforementioned four approaches to evaluate their performance for sparse biological networks by including bootstrap strategies. Specifically, we applied the nonparametric bootstrap, known as the Efron method, and the Bayesian bootstrap method due to the fact that in real data, the number of observations per genomic random variable is typically very limited. Thus, we overcame this limitation by augmenting sample sizes in the selection of the optimal model. We tested the accuracy of our results on two real and two simulated datasets.

Keywords: Graphical lasso algorithm, model selection methods, bootstrap methods, biological networks

9.1 Introduction

In the structure of protein-protein interaction networks, the estimation of the systems becomes challenging, and the best-fitted mathematical model, which can explain the biological activation, turns into a model selection problem because of the high-dimension of systems (p) with respect to the number of observations per gene (n), i.e., $n \ll p$, the sparsity of interactions, and the high correlation between genes. There are some classical model selection criteria, such as Akaike's information criterion (AIC)

Mehmet Ali Kaygusuz, Department of Economics, Anadolu University, 26470 Eskisehir, Turkey, https://orcid.org/0000-0002-4165-3791
Vilda Purutçuoğlu, Department of Statistics, Middle East Technical University, 06800 Ankara, Turkey, https://orcid.org/0000-0002-3913-9005

https://doi.org/10.1515/9783110717655-009

and Bayesian information criterion (BIC), besides some specific methods suggested more recently for sparse biological networks, like the extended BIC, stability selection and stability approach to regularization selection in the high-dimensional setting.

Basically, the most crucial part in the estimation of sparse biological networks is to infer the covariance matrix. It is one of the most important problems in classical statistics and modern statistics under high-dimensional data. Johnstone [1] and Ledoit and Wolf [2] have addressed this problem for the regularization of estimation under different cases, such as the principal component analysis and the high-dimensional covariance matrices. Bickel and Levina [3–6] have worked to improve the works of Johnstone [1] and Ledoit and Wolf [2] for high-dimensional data. Specifically, Cai et al. [5] computed the optimal rate of convergence to estimate the covariance matrix under different cases, and Ravikumar et al. [6] did not use any distributional assumptions, focusing instead on the tail behavior of the maximum deviation of the population and the sample covariance matrix.

Accordingly, as the general point of view in these calculations, it is accepted that the empirical covariance matrix comes from the normal (Gaussian) distribution. However, representing the sample size by n and the number of random variables by p, a random variable from the normal distribution with mean μ and covariance Σ_{pxp}, i.e., $N(\mu,\Sigma_{[}p_{)})$ does not have a consistent estimator of the covariance matrix, particularly when $p \gg n$. To solve this challenge, authors in [7–10] have suggested a faster algorithm for the graphical lasso model, which uses graphical methods to estimate a regularized covariance matrix. Later, Javanmard and Montanari [11] and Jancova and Van de Geer [12] have studied confidence intervals for high-dimensional covariance matrix. Jancova and Van de Geer [12] have implemented thresholding desparsified estimator for the low-dimensional parameters of the sparse precision matrix in order to select the edges in the graphical models. Moreover, Javanmard and Montanari [11] have applied the unbiased form of the regularized M-estimator for the high-dimensional regression and the covariance matrix. More details about high-dimensional statistics and covariance estimation can be found in the books by Bühlmann and Van de Geer [13] and Wainwright [14].

On the other hand, some nonparametric and semiparametric versions of the Gaussian graphical model have also been suggested, such as the semiparametric estimation of high-dimensional undirected graphs by Liu et al. [15], semiparametric Gaussian copula graphical models by Liu et al. [16] and the loop-based conic multivariate adaptive regression splines (CMARS) for sparse biological networks by Ayyıldız et al. [17]. Additionally, Yazıcı et al. [18] have used three bootstrap methods, including the wild bootstrap approach, to increase the computational efficiency of CMARS for small and medium-sized data.

Accordingly, Mohammadi and Wit [19] and Purutçuoğlu and Farnoidkia [20] have implemented the Bayesian techniques to estimate the model parameters and to compare the best-fitted model via binary accuracy measures such as F-measure and Matthew's correlation coefficient. Then, Williams et al. [21] performed the Bayesian model selection methods for the Gaussian graphical model. Furthermore, Epskamp et al. [22] and Williams et al. [23] have applied some bootstrap approaches to obtain better results for the model selection of psychological networks. Lastly, Kaygusuz and Purutçuoğlu [24] and Bülbül et al. [25] have shown that consistent AIC with Fisher information (CAICF) and information and complexity selection (ICOMP) approaches have better accuracies in sparse biological networks if they are used as the model selection criteria within the calculation of the Gaussian graphical model in the construction of underlying networks.

In this study, we aim to extend these analyses by proposing nonparametric and Bayesian bootstrap (BB) methods to improve the performance of model selection criteria under both low- and high-dimensional biological systems. The chapter is organized as follows: we introduce the graphical lasso algorithm and model selection methods, specifically AIC, BIC, CAICF and ICOMP, in Section 9.2. Then, we present the bootstrap methods, namely Efron's nonparametric method and the BB method in Sections 9.3 and 9.4, respectively. Later, we apply the bootstrapped model selection criteria to real and simulated datasets and evaluate their accuracies. Finally, in Section 9.5, we conclude our findings and discuss future work.

9.2 Graphical lasso algorithm and model selection criteria

In this section, we present the definition of the graphical lasso algorithm, which was proposed by Meinhausen and Bühlmann [7], and the model parameters estimated for high-dimensional data by Friedman et al. [8]. This section also includes the definition of classical model selection methods such as AIC and BIC, as well as the consistent AIC with Fisher information matrix (CAICF) and information and complexity (ICOMP) criterion, as extensions of those criteria.

9.2.1 Graphical lasso algorithm

The graphical lasso algorithm estimates the positive definite covariance matrix when we have n number of observations and p number of random variables. In our situation, we have the $p \gg n$ problem, meaning that the ordinary maximum likelihood estimators (MLEs) can be calculated, but they do not have closed forms. To address this problem, Yuan and Lin [10] has suggested the penalized maximum likelihood approach, and Friedman et al. [8] have developed the following equation as the graphical lasso algorithm:

$$\mathrm{logdet}(\Theta) - \mathrm{tr}(S\Theta) - \lambda \parallel \Theta \parallel_1 \tag{9.1}$$

where S is the empirical covariance matrix and $\Theta = \Sigma^{-1}$ denotes a non-negative precision matrix, while θij implies the corresponding variables between Xj and X_k, which are conditionally dependent. Furthermore, tr(·) shows the trace, and λ is a non-negative tuning parameter that controls the amount of l_1-shrinkage. Thus, this becomes a semidefinite programming problem for the variable Θ.

Banerjee et al. (2007) indicated that eq. (9.1) is convex and that σ can be estimated. Hence, they suggested a method for the problem of sparse maximum likelihood, which is based on the optimization over each row and the corresponding column of W in a block coordinate descent fashion. In this calculation, we can write the optimality conditions using the subgradient expression as follows:

$$\Theta^{-1} - S - \lambda\Gamma(\Theta) = 0 \tag{9.2}$$

where Γ is a matrix of the component signs of Θ. So, the graphical lasso algorithm uses the block coordinate method for solving W, S, Θ and Γ in eq. (9.2)

$$\Theta = \begin{pmatrix} \Theta 11 & \theta 12 \\ \theta 21 & \theta 22 \end{pmatrix} \Gamma = \begin{pmatrix} \Gamma 11 & \gamma 12 \\ \gamma 21 & \gamma 22 \end{pmatrix} \tag{9.3}$$

where Θ_{11} is a $(p-1) \times (p-1)$-dimensional matrix, θ_{12} is a $(p-1)$-dimensional vector and θ_{22} is a scalar value. Similarly, the matrices of W and S have been partitioned in the same way. Now, we can write $W = \Theta^{-1}$ as follows:

$$\begin{pmatrix} W11 & w12 \\ w21 & w22 \end{pmatrix} \begin{pmatrix} \Theta 11 & \theta 12 \\ \theta 21 & \theta 22 \end{pmatrix} = I \tag{9.4}$$

When we want to write the pth column, we can obtain the following equation:

$$w_{12} - s_{12} - \lambda\gamma_{12} = 0 \tag{9.5}$$

with $w12 = W11\beta$. This is

$$w_{11}\beta - s_{12} - \lambda\gamma_{12} = 0 \tag{9.6}$$

where the stationary equation for l_1-regularized quadratic program is given as follows for $\theta_{22} > 0$:

$$\min_{\beta \in \mathrm{R}} 0.5\beta_{w11}\beta + \beta_{s12} + \lambda \, || \, \beta||_1 \tag{9.7}$$

in which W_{11} is assumed to be fixed, as it is analogous to a lasso regression problem of the last variable on the rest, except the cross-product matrix S_{11} which is replaced by the current estimate W_{11} by exploiting the sparsity in β. This problem may be solved more efficiently by using element-wise coordinates as follows:

$$1/\hat{\theta}\ 22 = w22 - \widehat{\beta}\ ,\widehat{w}_{12} \tag{9.8}$$

Finally, $\hat{\theta}_{12}$ might be recovered from $\widehat{\beta}$ and $\hat{\theta}_{22}$. The pseudoform of the algorithm is represented as follows:

1. Initialization $\mathbf{W} = S + \lambda_i$. The diagonal includes a constant.
2. Repeat for $j = 1, 2, \ldots, p$, until convergence:
   (i) Partition the matrix **W** into two parts: part 1 includes all but the jth row and column, and part 2 includes the jth row and column.
   (ii) Solve the estimation equations $\mathbf{W}_{11}\beta - s_{12} + \lambda \times \text{sign}(\beta) = 0$ by using the cyclical coordinate algorithm for the modified lasso.
   (iii) Update $w12 = 11\widehat{\beta}$.
3. In the final cycle (for each j), solve for $\hat{\theta}12 = -\beta \times \hat{\theta}22$ with $1/\hat{\theta}22 = w22 - wT\widehat{\beta}$.

9.2.2 Classical model selection methods: Akaike's information criterion and Bayesian information criterion

AIC, proposed by Akaike [26], is a penalized version of the attained maximum log-likelihood for each candidate model. On the other hand, Schwarz's BIC [27] also has the form of a penalized log-likelihood function. However, BIC imposes a stronger penalty for additional parameters compared to AIC. Furthermore, BIC can be more consistent than AIC. In these criteria, it is assumed that the data are generated from the true density function $f_{\theta 0}$ with a parameter vector. $\theta_0 = (\theta_0, \ldots, \theta d)T$.

Hence, Akaike [26, 28] has proposed choosing a model that minimizes the Kullback-Leibler (KL) distance of the fitted model from the true model. He considers the MLE of the parameter vector M and provides a proof with an additive constant. Thus, the estimated KL distance might be asymptotically adopted as

$$\text{AIC}(M) = -2\ \ln(M) - 2\text{dim}(M) \tag{9.9}$$

where ln(M) is the log-likelihood function, and dim(·) indicates the dimension of the model. Regarding the expression in eq. (9.9), the highest score of AIC indicates the best-fitted model.

Besides AIC, Schwarz [27] has suggested a Bayesian approach that includes a prior distribution on the candidate models along with a prior distribution on the parameter vector in the models:

$$\text{BIC}(M) = -2\ \ln(M) - (\log n)\text{dim}(M) \tag{9.10}$$

in which the number of parameters, dim(M), is estimated from the candidate model, and n refers to the data sample size. Thus, the best model is found via the maximum score according to eq. (9.10).

On the other hand, as a disadvantage of both AIC and BIC, when the number of parameters is much higher than the number of observations, AIC and BIC might be inconsistent. For this reason, some extensions of classical AIC and BIC have been proposed [29, 30]. In the following section, we present some extensions of AIC proposed by Bozdoğan [30, 31].

9.2.3 Extensions of Akaike's information criterion: CAICF and ICOMP

In the literature, some extensions of AIC are used to select the right model more consistently with additional penalty terms. Three of these extended selection criteria are proposed by Bozdoğan [30, 31] which is originated from Kullback information and the Fisher information matrix. Bozdoğan [30, 31] suggests that selection methods can calculate a metric distance between the model and the true distribution, and a consistent inference method should make this distance as small as possible. In the study of Bozdoğan [31], the Kullback-Leibler divergence is applied, and this novel criterion is called the consistent AIC (CAIC), as follows:

$$\mathrm{CAIC}(k) = -2\log L\left(\hat{\theta}_k\right) + k[\log n + 1] \tag{9.11}$$

where the likelihood of θ is denoted by $\log L(\hat{\theta}_k)$ and k shows the degrees of freedom of the distribution. In fact, CAIC(k) is similar to BIC of k log n and k[log $n + 1$] terms that have a stronger penalty term.

Another Bozdoğan's selection method is based on the consistent AIC with the Fisher information CAICF(k) matrix. This method has an increasing penalty term for overparameterization, which can be represented as follows:

$$\mathrm{CAICF(k)} = -2\log \mathrm{L}\left(\hat{\theta}_k\right) + \mathrm{k}[\log \mathrm{n} + 2] + \log\left|\mathrm{I}\left(\hat{\theta}_k\right)\right| \tag{9.12}$$

in which $\log L(\hat{\theta}_k)$ denotes the likelihood estimation of θ, as used beforehand, k is the degrees of freedom of the distribution, and $I(\hat{\theta}_k)\hat{F}^{-1}$ presents the Fisher information matrix.

Lastly, another extension of the CAIC model selection method is known as the ICOMP method, which was proposed by Bozdoğan [30]. This new criterion has the ability to penalize the free parameters and the covariance matrix directly, as follows:

$$\mathrm{ICOMP} = -2\log L\left(\hat{\theta}_k\right) + 2C(\hat{\Sigma}) \qquad [9.13]$$

where $\log L(\hat{\theta}_k)$ is the log-likelihood function, $\hat{\theta}_k$ indicates the maximum likelihood estimate of the parameter vector of θ_k, C represents a real-valued complexity measure, and $\hat{\Sigma} = \widehat{cov}(\hat{\theta}_k)$ shows the estimated covariance matrix of the parameter vector of the candidate model. This covariance matrix might be obtained in different ways. Bozdoğan's [30] choice is the computation of the inverse of the Cramer-Rao lower bound matrix, which is the function of the estimated inverse Fisher information matrix (IFIM) of the model as follows:

$$\hat{F}^{-1} = \left\{ -E\left(\frac{\partial^2 \log L(\theta)}{\partial\theta\partial\theta'}\right)\right\}^{-1} \tag{9.14}$$

In this expression, the $(s \times s)$-dimensional second-order partial derivatives of the log-likelihood function of the estimated model are denoted by $\hat{F}^{-1}$ and ∂, which stands for the partial derivative of the given function. Finally, a more general form of ICOMP can be expressed by the following equation:

$$\text{ICOMP} = -2\log L\left(\hat{\theta}_k\right) + 2C\left(\hat{F}^{-1}\right) \tag{9.15}$$

when

$$C\left(\hat{F}^{-1}\right) = \frac{s}{2}\log\left[\frac{\widehat{\text{tr}F}^{-1}}{s}\right] - \frac{1}{2}\log\left|\hat{F}^{-1}\right| \tag{9.16}$$

Equation (9.16) represents the information complexity of the estimated IFIM of the model, and $s = \dim(\hat{F}^{-1}) = \text{rank}(\hat{F}^{-1})$ while $\dim(\cdot)$ denotes the dimension of the given matrix as used previously.

9.3 Bootstrap methods

9.3.1 Parametric and nonparametric bootstrap methods

The bootstrap methods allow the estimation of the sampling distribution of statistics by using random sampling methods. Basically, the bootstrap methods involve resampling iterations, and it is shown that the bootstrap of [32] is a powerful nonparametric tool for approximating the sampling distribution and the variance of the statistic.

In our analysis, we assume that the data are independent and identically distributed (iid). On the other hand, the most common choice for approximating the distribution is the empirical distribution function of the observed data. For more details about different bootstrap procedures under distinct statistical models, the monographs by [33–35] can be suggested. Regarding bootstrap methods, it is given that a specific random variable $R(X, F)$ depends on both X and the unknown distribution F, and the aim is to estimate the sampling distribution of R which is based on observations of data x. In the related literature, the conservative jackknife theory focuses on two-parameter choices of R, such as the mean, correlation and standard deviation of F, and $T(X)$ is taken as the estimator of $\theta(F)$, whose steps for the one-sample case can be listed as follows:

Step 1: Build the sample probability distribution $\hat{F}$ which divides n (the number of total observations) at each point $x_1, x_2, x_3, \ldots, x_n$.

Step 2: When $\hat{F}$ is constant, draw a random sample of size n from $\hat{F}$ via $X_i^* = x_i^*$, $X_i^* = \hat{F}$, $i = 1, 2, \ldots, n$ in such a way that the bootstrap sample can be called as $X^* = \left(X_1^*, X_2^*, \ldots, X_n^*\right)$ while $X^* = \left(x_1^*, x_2^*, \ldots, x_n^*\right)$.

Step 3: Approximate the sampling distribution of $R(X, F)$ using the bootstrap distribution:

$$R^* = R\left(X^*, \hat{F}\right) \tag{9.17}$$

In eq. (9.17), if $F = F^*$, the distribution of R^* can be calculated theoretically when the data are observed, and if it equals the distribution of R. Therefore, a nonparametric estimator of the distribution of R is one of the most accurate estimators without any prior assumption of F as long as $F = \hat{F}$. Thus, one of the simplest examples of the bootstrap methods can be seen as the binomial distribution when the values of the distribution of F are 0 and 1, as follows:

$$R(X, F) = \overline{X} - \theta\left(F, \overline{X}\right) = \sum_{i=1}^{n} (\ Xi)\ /n \tag{9.18}$$

In eq. (9.18), the observations $X = x$ and the bootstrap samples $X^* = \left(X_1^*, X_2^*, \ldots, X_n^*\right)$ have an independent component that equals one with a probability $\overline{x} = \theta(\hat{F})$ and zero with a probability $(1 - \overline{x})$. We can show that the associated binomial results are as follows:

$$R^* = R\left(X^*, \hat{F}\right) = \overline{X}^* - \overline{x} \tag{9.19}$$

$$E^* \left(\overline{X}^* - \overline{x}\right) = 0,\ \mathrm{Var}^* \left(\overline{X}^* - \overline{x}\right) = \overline{x}\ (1 - \overline{x})/n \tag{9.20}$$

In the above expressions, $\overline{X}$ is an unbiased estimator of θ and has a variance similar to $\overline{x}\ (1 - \overline{x}\ )/n$. Accordingly, the underlying procedure can be summarized as follows: The dataset consists of an iid sample of size n from an unknown probability distribution F on the real line via

$$(X_1, \ldots, X_n) \sim F \tag{9.21}$$

whose observed values are denoted by $X_1 = x_1,\ X_2 = x_2, \ldots,\ X_n = x_n$. Thus, we need to calculate the mean of the sample $\overline{x} = x_{i/n}$ by using an estimate of the expectation F as shown in eq. (9.21). Hence, we can compute the standard deviation of the sample by

$$\hat{\sigma}^2 = \left[\frac{1}{n(n-1)} \sum_{i=1}^{n}\ (xi - \overline{x})2\right] \tag{9.22}$$

In eq. (9.22), $\hat{\sigma}$ is the standard deviation of $\overline{X}$ and the root indicates the square error of the estimation. Thus, let $\hat{F}$ be the distribution of the data. Each x_i has a $1/n$ probability as stated before. Considering that $X_1^*, X_2^*, \ldots, X_n^*$ are iid samples from $\hat{F}$,

$$\left(X_1^*, X_2^*, \ldots, X_n^*\right) \sim \hat{F}. \tag{9.23}$$

In eq. (9.23), X_i^* is a random sample drawn with replacement from observed values $x_1, x_2, \ldots, x_n$. Therefore, $\overline{X}^* = X_i^* / n$ has a variance as follows:

$$\text{Var}^*\left(\overline{X}^*\right) = \sum_{i=1}^{n} (x_i - \overline{x})2 \tag{9.24}$$

where Var(·) shows the variance of the distribution of simulated data $\hat{F}$. Hereby, the bootstrap estimate of the standard deviation of an estimator can be denoted by $\hat{\theta}$ $(X_1, \ldots, X_n)$.

As a result, the steps of the parametric bootstrap can be listed as follows:

Step 1: $X_i(i = 1, \ldots, n)$ are iid from$f(,;\theta)\theta \in \Theta, \theta_0$ is used to show the true value of the parameter.

Step 2: $\hat{\theta}_n = h(X_1, \ldots, X_n)$ is an estimator of θ and Tn is a standardized random variable $\hat{T}_n = n^{1/2}(\hat{\theta}_n - \theta_0)/v_n$, where vn is the standard deviation of $v_n{}^2 = h(X_1, \ldots, X_n)$.

Step 3: Consider X_i^*being iid from $f\left(,;\hat{\theta}\right)$ and we apply the formula $\hat{\theta}$ and v_n while $T_n^* = n^{\frac{1}{2}}\left(\widehat{\theta_n^*.} - \hat{\theta}obs\right)/v_n^*$.

Step 4: Use M Monte Carlo simulations, where each run is composed of samples from the approximate sampling distribution.

On the other hand, the nonparametric bootstrap procedure can be summarized as follows:

Step 1: $X_i(i = 1, \ldots, n)$ are iid from $f[\cdot, \theta)$.

Step 2: $\hat{\theta}_n = h(X_1, \ldots, X_n)$ is an estimator of θ and T_n is the standardized random variable $\hat{T}_n = n^{1/2}(\hat{\theta}_n - \theta_0)/v_n$, where vn is the standard deviation of $v_n^2 = (X_1, \ldots, X_n)$.

Step 3: Consider X_i^* be being iid from Fn which is *the empirical distribution function* of $(X_1, \ldots, X_n)$. Here, we use the formula for $\hat{\theta}_n$ and v_n $\widehat{\theta_n^*} = h(X_1^*, \ldots, X_n^*)$ $(v_n^*)^2 = h(X_1^*, \ldots, X_n^*)$ while $T_n^* = n^{1/2}\left(\widehat{\theta_n^*.} - \hat{\theta}_{obs}\right)/v_n^*$.

Step 4: Use M Monte Carlo simulations, where each run is composed of n observations, which are taken from the approximate sampling distribution of $\hat{\theta}_n$.

9.3.2 The Bayesian bootstrap

When the classical bootstrap approach makes the simulation of the sampling distribution of a statistical parameter, the BB method, suggested by Rubin [36], uses the posterior distribution of the parameter to simulate the data. In this method, every BB iteration generates a posterior distribution for each x_i. In the posterior probability of each n, x has a $1/n$ probability, but it also has some variability.

Accordingly, the BB iteration can generate data from a uniform distribution U [0,1] with random variables $u_1, u_2, \ldots, u_n$ and calculate the difference between them, such as

$$u_2 - u_1, u_3 - u_2, \ldots, u_n - u_{n-1}, \tag{9.25}$$

where $u_n = 1$ and $u_0 = 0$. Thus, the BB simulations provide us with the BB distribution of X and any parameter of this distribution.

The main difference between the Efron bootstrap approach and the BB method is how we assign probability weights to x_i in simulations. Indeed, both methods are similar, but the interpretations of the resulting distributions can differ. The BB samples are drawn from the posterior distribution of the parameter θ while the bootstrap method samples from the estimated sampling distribution of a statistic θ to make $\hat{\theta}$ inferences. Therefore, the BB method has an obvious advantage over the parametric bootstrap method due to the results of the inference of parameters.

Rubin [36] also implements a noninformative prior by combining the BB approach with a multinomial likelihood Z_i, leading to a Dirichlet (1, . . ., 1) posterior distribution of w_i. Hence, the posterior distribution of $T(F)$ is estimated by the Monte Carlo methods, which generate $w_{(b)}$ from a Dirichlet (1, . . ., 1) distribution and calculate $T(F_{(b)})$ from each sample $w_{(b)}$ as follows:

$$\pi(\theta 1, \ldots, \theta J) = \Gamma\left(\sum_{j=1}^{n} aj\right) \Big/ \left(\prod_{j=1}^{n} \Gamma(aj)\right) \prod_{j=1}^{n} \Gamma\left(\theta^{aj-1}\right) \tag{9.26}$$

Accordingly, $f(\theta) \in \Theta = S^{j-1}$ and zero otherwise. The indices of the parameters of the Dirichlet distribution $a = (a_1, \ldots, a_j)' \pi$ show the vector of positive real numbers. So, conditional on θ, the data having a multinomial distribution can be presented as follows:

$$f(z|\theta) = N!/N1! \cdots NJ! \prod_{j=1}^{n} \Gamma\left(\theta_j^{aj-1}\right) \tag{9.27}$$

In eq. (9.27), we update our results by using Bayes's rule about the population distribution. The conditional distribution of θ given $\mathbf{Z} = \mathbf{z}$ is called the posterior distribution, which is expressed as follows:

$$\bar{\pi}(\theta|z) = f(z|\theta)\pi(\theta)/f(z|\theta)\pi(\theta)d\theta \tag{9.28}$$

With the multinomial likelihood and Dirichlet prior,

$$\bar{\pi}(\theta|z) = \prod_{j=1}^{n} \Gamma\left(\theta j^{Nj+aj-1}\right) / \int_{Sj-1} \prod_{j=1}^{n} \Gamma\left(\theta j^{Nj+aj-1}\right) d\theta 1 \cdots d\theta J \tag{9.29}$$

Then, by using the Beta function, we can write the conditional distribution as follows:

$$\int_{Sj-1} \prod_{j=1}^{n} \Gamma\left(\theta j^{Nj+aj-1}\right) d\theta 1 \cdots d\theta J = \left(\prod_{j=1}^{n} \Gamma(Nj+aj)\right) \Big/ \Gamma\left(\sum_{j=1}^{n} Nj+aj\right) \tag{9.30}$$

Now, our posterior distribution can be described as a part of the Dirichlet distribution:

$$\bar{\pi}(\theta|z) = \left[\Gamma\left(\sum_{j=1}^{n} Nj+aj\right) \Big/ \prod_{j=1}^{n} \Gamma(Nj+aj)\right] \prod_{j=1}^{J} \Gamma\left(\theta j^{Nj+aj-1}\right) \tag{9.31}$$

where the conjugate distribution of the multinomial distribution is known as the Dirichlet distribution. Thus, we assume that $\overline{a}j = aj + Nj$ for $j = 1, \ldots, J$, $\overline{a}0 = \sum_{j=1}^{n} \overline{a}j$ and $\overline{\theta} = ((aj+Nj)/(\sum_{j=1}^{n} Nj+aj)) = \overline{a}j/(\overline{a}0)$ Therefore, the posterior mean of θ can be shown via

$$E[\theta|\mathbf{Z} = \mathbf{z}; a] = \overline{\theta} \tag{9.32}$$

9.4 Application with real data

9.4.1 Real data description

In this section, we use two real benchmark datasets, whose descriptions are presented as follows:

i) Data 1: It is from the PGC-1A pathway, peroxisome proliferator-activated receptor gamma co-activator-1 alpha, which is a tissue-specific coactivator and coordinates transcriptional programs crucial for energy metabolism and energy homeostasis. The disrupted increases in the PGC-1A activity, which leads to many pathological conditions such as heart failure and diabetes. The NCI-60 cell line (p53) datasets are composed of the gene expression profiling of 33-p53-mutated cancer cell lines. We have studied 20 genes from this pathway taken from the study of Rahmatallah et al. [37].
ii) Data 1: It is called the Human Gene Expression Data, which involves the gene expression of B-lymphocyte cells from Utah residents with Northern and Western European ancestry samples. The genes from 60 unrelated individuals are investigated for 100 different transcripts, taken from the study of Bahadra and Mallick [38] and Chen and Chen [39].

9.4.2 Accuracy measures

In the analyses, we use the accuracy value, precision, and *F*-measure for two real and two simulated datasets. These measures are calculated according to four values listed as true positive (TP), true negative (TN), false positive (FP) and false negative (FN). The TP indicates the number of correctly classified objects that have positive labels and the

FP is the number of misclassified objects that have positive labels. Furthermore, the TN shows the number of correctly classified objects that have no link, as denoted by 0. Here, the perfect accuracy is obtained when the precision equals to 9. Finally, the FN indicates the number of misclassified objects that have a positive label. This information may be constructed as a confusion matrix, as presented in Table 9.1.

Table 9.1: Accuracy measures.

	Actual class	
Predicted class	**Positive**	**Negative**
Positive	TP	FP
Negative	FN	TN

Here, the accuracy measure is defined as the ratio of correctly classified objects in both labels to the total number of classified objects, as follows:

$$\textbf{Accuracy} = (\mathrm{TP} + \mathrm{TN}) \,/\, (\mathrm{TP} + \mathrm{FP} + \mathrm{TN} + \mathrm{FN}) \tag{9.33}$$

On the other hand, we can define the *F*-measure as the harmonic mean of precision and recall:

$$F\text{-measure} = (2\mathrm{TP}) \,/\, (2\mathrm{TP} + \mathrm{FP} + \mathrm{FN}) \tag{9.34}$$

Finally, the precision measure can be computed as follows:

$$\textbf{Precision} = (\mathrm{TP})/(\mathrm{TP} + \mathrm{FP}) \tag{9.35}$$

9.4.3 Results

We primarily perform ICOMP and CAICF selection methods to evaluate the performance with and without bootstraps in high dimensions. Accordingly, we analyze the gene expression and cancer cells of p53 datasets. Furthermore, we simulate two network scenarios for different sample sizes and random variables under scale-free and random network topologies that indicate distinct network sparsity and have a multivariate normal distribution. Hence, we take 50 and 100 as the numbers of genes (p) when the number of observations (n) is 10 (Figure 9.1). Additionally, in these analyses, we have set the simulation number to 100, the number of bootstrap iterations (B) to 1,000 and the number of posterior distributions for the BB to 4,000 for the simulated data and real datasets.

The results in Tables 9.2 and 9.3 indicate that the bootstrap under parametric and Bayesian approaches improves the accuracy of estimates in the CAICF model selection criterion across all datasets. On the other hand, ICOMP is irrelevant to the bootstrap algorithm.

Table 9.2: The results of nonparametric bootstrap approaches with and without using accuracy (acc), precision (pre), and F-measure (F) for sample sizes per gene (n) and the total number of genes (p).

Pathway names	Measures	ICOMP (without)	ICOMP (with)	CAICF (with- out)	CAICF (with)
1. PGC-1A	acc	0.815	0.815	**0.670**	**0.815**
pathway	pre	1.00	1.000	**0.400**	**1.000**
$n = 33$ $p = 20$	F	0.351	0.351	**0.535**	**0.351**
2. Human gene	acc	0.991	0.991	**0.932**	**0.991**
expression.	pre	1.000	1.000	**0.214**	**1.000**
$n = 60$ $p = 100$	F	0.690	0.690	**0.351**	**0.690**
3. Simulated scale-	acc	0.980	0.980	**0.822**	**0.980**
free network	pre	1.000	1.000	**0.083**	**1.000**
$n = 10$ $p = 100$	F	0.503	0.503	**0.142**	**0.503**
4. Simulated	acc	0.938	0.9384	**0.716**	**0.946**
random network	pre	1.000	1.000	**0.142**	**1.000**
$n = 10$ $p = 50$	F	0.394	0.394	**0.227**	**0.424**

Table 9.3: The results of Bayesian bootstrap approaches with and without accuracy (acc), precision (pre), and F-measure (F) for sample sizes per gene (n) and the total number of genes (p).

Pathway names	Measures	ICOMP (without)	ICOMP (with)	CAICF (with out)	CAICF (with)
1. PGC-1A	acc	0.815	0.815	**0.670**	**0.815**
pathway	pre	1.000	1.000	**0.400**	**1.000**
$n = 33, p = 20$	F	0.351	0.351	**0.535**	**0.351**
2. Human gene	acc	0.991	0.991	**0.932**	**0.991**
exp	Pre	1.000	1.000	**0.214**	**1.000**
$n = 60, p = 100$	F	0.690	0.690	**0.350**	**0.690**
3. Simulated	acc	0.948	0.948	**0.712**	**0.953**
random network.	pre	1.000	1.000	**0.143**	**1.000**
$n = 10, p = 50$	F	0.435	0.435	**0.234**	**0.459**
4. Simulated	acc	0.980	0.980	**0.809**	**0.980**
scale-free	pre	1.000	1.000	**0.072**	**1.000**
network	F	0.503	0.503	**0.125**	**0.503**
$n = 10, p = 100$					

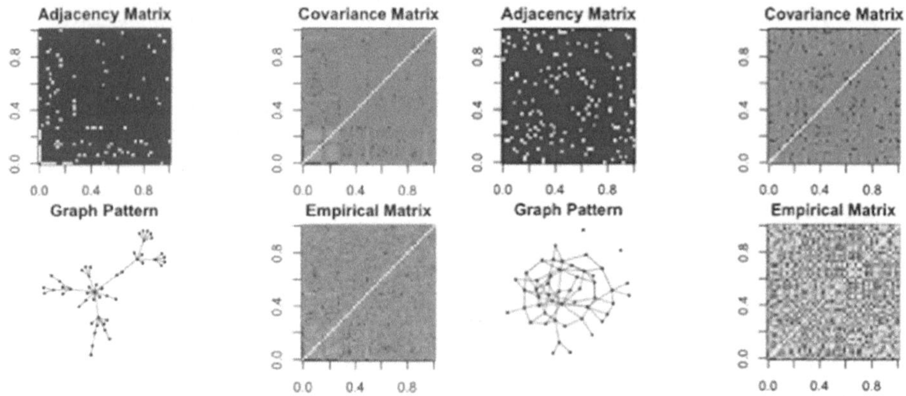

(a) Scale-free network $n=10$, $p=50$ (b) Random network $n=10$, $p=100$

Figure 9.1: The graphical representation of the simulated network under (a) scale-free network structure with $n = 10$ and $p = 50$, and (b) random network structure with $n = 10$ and $p = 100$.

9.5 Conclusion

In this study, we compared CAICF and ICOMP selection methods and used nonparametric and BB methods to improve the accuracy of selection methods when the number of random variables (p) is much greater than the number of observations (n). We applied two real datasets for this purpose: the PGC-1A pathway, which has 33 cancer cells and 20 genes, and the Human gene expression data, which comprises 100 transcripts from 60 individuals. The findings showed that CAICF converges to ICOMP after the bootstrap step. Moreover, the bootstrap methods improved the accuracy of the selection methods when we compared the CAICF and ICOMP selection methods without the bootstrap estimation in the high-dimensional setting. We also evaluated the performances under simulated pathways generated under scale-free and random network structures when the number of genes ($p = 50, 100$) is higher than the number of observations ($n = 10$). From the results, it can be seen that in scale-free networks, CAICF converges to ICOMP after the bootstraps, but in random networks, CAICF is better than ICOMP after the bootstrap estimation. Finally, we consider that these outcomes are promising, but more comprehensive analyses can be done under distinct sample sizes (n) and the number of random variables (p) to obtain a more detailed assessment of the effect of the bootstrap in distinct model selection criteria.

References

[1] I. M. Johnstone. On the distribution of the largest eigenvalue in principal components analysis. Annals of Statistics, 29:295–327, 2001.

[2] O. Ledoit and M. Wolf. A well-conditioned estimator for large-dimensional covariance matrices. Journal of multivariate analysis, 88:365–411, 2003.

[3] P. J. Bickel and E. Levina. Regularized estimation of large covariance matrices. Annals of statistics, 36:199–227, 2008a.

[4] P. J. Bickel and E. Levina. Covariance regularization by thresholding. Annals of statistics, 36:2577–2604, 2008b.

[5] T. T. Cai, C. H. Zhang and H. H. Zhou. Optimal rate of convergence for covariance matrix estimation. Annals of Statistics, 38:2118–2144, 2010 .

[6] P. Ravikumar, M. J. Wainwright, G. Raskutti and B. Yu. High-dimensional covariance estimation: Convergence rates of l1-regularized log-determinant divergence. Electronic Journal of Statistics, 5:935–980, 2011 .

[7] N. Meinhausen and P. Bühlmann. High dimensional graphs and variable selection with lasso. Annals of Statistics, 34:1436–1462, 2006.

[8] J. Friedman, T. Hastie and R. Tibshirani. Sparse inverse covariance estimation with the graphical lasso. Biostatistics, 9:432–441, 2007.

[9] A. D'aspremont, O. Banerjee and L. El Ghaoui. First-order methods for sparse covariance selection. SIAM Journal on Matrix Analysis and Applications, 2007 .

[10] M. Yuan and Y. Lin. Model selection and estimation in Gaussian graphical model. Biometrika, 94:19–35, 2007.

[11] A. Javanmard and A. Montanari. Confidence intervals and hypothesis testing for High-dimensional regression. Journal of Machine Learning Research, 15:2869–2909, 2014.

[12] J. Jancova and S. Van de Geer. Confidence intervals for High-dimensional in- verse covariance estimation. Electronic Journal of Statistics, 9:1205–1229, 2015.

[13] P. Bühlmann and S. Van de Geer. Statistics for High Dimensional Data: Methods, Theory, Application, Springer, 2011.

[14] M. J. Wainwright. High-Dimensional Statistics: A Non-Asymptotic Viewpoint, (Cambridge Series in Statistical and Probabilistic Mathematics), Cambridge University Press, 2019.

[15] H. Liu, J. Lafferty and L. Wasserman. The non-paranormal: Semiparametric estimation of high dimensional undirected graphs. Journal of Machine Learning Research. 10:2295–2328, 2009.

[16] H. Liu, J. Lafferty and L. Wasserman. Sparse Non-parametric Graphical Models. Statist, 2012 .

[17] E. Ayyıldız, V. Purutçuoğlu„ and G. W. Weber. Loop-based conic multivariate adaptive regression splines is a novel method for advanced construction of complex biological networks. European Journal of Operational Research, 3:852–861, 2018 .

[18] C. Yazıcı, F. Yerlikaya-Özkurt and I. Batmaz. A computational approach to non-parametric regression: Bootstrapping CMARS method. Machine Learning, 101:211–230, 2015.

[19] A. Mohammadi and E. C. Wit. Bayesian structure learning in sparse Gaussian graphical models. Bayesian Analysis, 10(1):109–138, 2015.

[20] V. Purutçuoğlu and H. Farnoidkia. Gibbs sampling in inference copula Gaussian graphical models adapted to biological networks. Acta Physica Polonika, 132:3, 2017.

[21] R. D. Williams, J. Piironen, A. Vehtari and P. Rast. Bayesian estimation of Gaussian graphical models with predictive covariance selection, Arxiv, 2018b.

[22] S. Epskamp, D. Borsboom and E. I. Fried. Estimating psychological networks and their accuracy: A tutorial paper. Behavior Research 50:195–210, 2018.

[23] D. R. Williams, M. Rhemtulla, A. C. Wysocki and P. Rast. on non-regularized estimation of psychological networks, Arxiv, 2018a.

[24] M. A. Kaygusuz and V. Purutçuoğlu. Model Selection Methods for Sparse Biological Networks Chapter. In: T. Yiğit, U. Köse Editors. Lecture Notes on Data Engineering and Communications Technologies. Springer, 2019.
[25] G. B. Bülbül, V. Purutçuoğlu and E. Purutçuoğlu. Novel model selection criteria on sparse biological networks. International Journal of Environmental Science and Technology, 16:1–12, 2019 .
[26] H. Akaike. Information Theory and an Extension of the Maximum Likelihood Principle. In: B. N. Petrov and F. Csaki Editors. Second International Symposium on Information Theory. Budapest: Akademiai Kiad, 267–281, 1973.
[27] G. Schwarz. Estimating the dimension of a model. Annals of Statistics, 6:461–464, 1978.
[28] H. Akaike. A new look at the statistical model identification, IEEE Transaction Autocontrol, 19:716–723, 1974.
[29] R. Foygel and M. Drton. Extended Bayesian information criteria for Gaussian Graphical models. Advances in Neural Information Processing Systems, 23:2020–2028, 2010.
[30] H. Bozdoğan. A new class of information complexity (ICOMP) criteria with an application to costumer profiling and segmentation. 39(2):370–398, 2010.
[31] H. Bozdoğan. Model selection and AIC: The general theory and its analytical extensions. Pscychometrica, 52(3):345–370, 1987.
[32] B. Efron. Bootstrap methods: Another look at jackknife, Annals of Statistics, 7:1–26, 1979.
[33] B. Efron. The jackknife, the bootstrap and other re-sampling plans, SIAM, [Society for Industrial and Applied Mathematics, 1982.
[34] B. Efron and R. J. Tibshirani. An Introduction to the Bootstrap, CRC, 1993.
[35] A. C. Davison and D. V. Hinkley. Bootstrap Methods and Their Applications. In: Cambridge Series in Statistical and Probabilistic Mathematics, Cambridge University Press, 1997.
[36] D. B. Rubin. The Bayesian Bootstrap, The Annals of Statistics, 9:1130–1134, 1981.
[37] Y. Rahmatallah, F. Emmert-Streib and G. Glazko. Gene sets net correlation analysis (GSNCA): A multivariate gene coexpression test for gene tests. Bioinformatics, 30(3):360–388, 2014.
[38] A. Bahadra and B. K. Mallick. Joint high-dimensional Bayesian variable and co- variance selection with an application eQTL analysis, Biometrics, 69(2):447–457, 2013.
[39] J. Chen and Z. Chen. Extended Bayesian information criterion for model selection with large model space, Biometrika, 95:759–771, 2008.
[40] N. Meinhausen and P. Bühlmann. Stability selection, Journal of Royal Statistical Society, 72:417–473, 2010.
[41] O. Banerjee, L. El Ghaoui and L. D'Aspremont. Model selection through sparse maximum likelihood estimation for multivariate Gaussian or binary data, Journal of Machine Learning Research, 9:485–516, 2008.
[42] P. Zhao and B. Yu. On model selection consistency of lasso, Journal of Machine Learning Research, 7:2541–2563, 2006.
[43] D. N. Politis, J. P. Romano and M. Wolf. Subsampling, (Springer Series in Statistics), Springer, 1999.
[44] S. Boyd and L. Vanderberghe. Convex Optimization, Cambridge University Press, 2004.
[45] T. Hastie, R. Tibshirani and J. Friedman. The Elements of Statistical Learning: Data Mining, Inference and Prediction, Springer, New York, 2009.
[46] H. Liu, K. Roeder and L. Wasserman. Stability approach to regularization selection (StARS) for high dimensional graphical models. In: Proceeding of the twenty-third Annual conference on Neural Information Processing system (NIPS), 1–14, 2010.
[47] I. M. Johnstone and A. Y. Lu. On consistency and sparsity for principal components analysis in high dimensions. Journal of the American Statistical Association, 104:682–693, 2009.
[48] A. Dempster. Covariance selection, Biometrics, 28:157–175, 1972.
[49] G. Claeskans and N. L. Hjort. Model Selection and Model Averaging. In: Cambridge Series in Statistical and Probabilistic Mathematics, Cambridge University Press, 2008.

Cinzia Bellone, Pietro Ranucci and Valina Geropanta

Chapter 10 Technocracy in Governance: new directions in city functioning and urban planning

Abstract: In 2018, we published a chapter in the book *Intelligent Computing and Optimization* about how smart city tools, solutions and applications alter the contemporary notion of governance and lead to innovative methodologies for territorial planning [1]. We now come back to show the progress that has been made during the past 3 years in the sector, and see how these tools have really transformed the governance processes in cities where they are applied. Our new review of literature reveals how the processes of production/realization of spatial configurations are changing rapidly, both for the peculiarities of the tools that are used today for their management and for the new social and urban realities. The tools are now temporally and structurally unsuitable for reacting promptly (and in a sustainable way) to the economic and spatial needs of the population.

Keywords: Urban planning, technocracy, governance, and smart city

10.1 Introduction

The most common definition of a smart city states that investments in physical (transport) and intangible (ICT) communication infrastructures, linked to human and social capital, are the most important aspect of technological innovation, while quality of life and *sustainable* development are essential for the "construction" of the city. It is well known that the supporters of this view tend to consider the smart city through the concept of "efficiency" of the urban "machine" and the use of human capital.

Nonetheless, a new process of urban "construction" ("training and adaptation" of the spaces and functions of the city over time) could and should aim at something else. Instead of disqualifying the above-mentioned hypothesis, it should subordinate it to more demanding objectives that can be found in an increasingly better-defined relationship between *decision-makers and citizens,* with an explicit reference to such topics as "democracy, technology, technocracy." It is the argument of this contribution.

For this reason, the debate on these keywords must be directed toward a research path that identifies interrelated systems of relationships, because each of them alone

Cinzia Bellone, Guglielmo Marconi University, Rome, Italy, e-mail: c.bellone@unimarconi.it
Pietro Ranucci, Roma Tre University, Rome, Italy
Valina Geropanta, Greece Founder of Avrions, Technical University of Crete, Chania, Greece

https://doi.org/10.1515/9783110717655-010

is unable to control a process of urban development that generates sustainable social and economic paradigms and a definition of urban spaces and their operation in line with the desired living standards.

With reference to Italy and within the Triennial Plan of AgID[1] 2019–2021, something seems to be moving beyond the numerous smart city initiatives. See, for example, the remarkable program of digitization of the public administration (PA) within the PNRR (National Program of Recovery and Resilience) that draws from the 3-year plan (2019–2021) and addresses the digitization of the PA through developing digital infrastructure, facilitating cloud migration, supplying data and interoperability, digital services and digital citizenship, ensuring cybersecurity, digitization of large central administrations, basic digital skills and ICT purchasing process.

But, above all, in the Agid Triennial Plan, see the proposal for a National Plan for Intelligent Communities (art. 20, law 221/2012, by converse. Dl 179/2012), which proposes to expand the smart city concept beyond the "walls" of the city, invest in the territory and apply to it (Smart Landscape Platform) the system of citizen-manager relationships that smart city technology has put into action in order to allow citizens and local administrations to establish jointly the quality and forms of provision of urban and territorial services through transparent forms of collection and management of sensitive data of the citizens concerned:

> This concept changes the approach to the development of intelligent solutions, as it starts from the assumption that people and territory form a network of transversal connections to all areas of intervention combined with the Smart City, from the environment to the economy. In this new paradigm of territorial development, transport infrastructures, the role of enterprises and productivity and the collaboration between public and private entities acquires great importance. (Gaia Mussi, 22–01–20)

In connection with the above, see the Enea[2] Convergence Smart City and Community initiative. Aiming "to move from words to action," it represents a search that started before the approval of PNRR and aims for "the digital reorganization of the management processes of urban and territorial contexts, starting from conceptual, methodological, and technological convergence."

Although limited to the promotion of smart cities, the ENEA initiative seems to be a prerequisite for the PNRR precisely because it states that, in order to support national planning on these issues, it is necessary to start by sharing common languages to identify and operate on strategic data to be used for the efficiency of systems, particularly urban ones. The initiative concludes that, if this happens, it would foster col-

1 Agid – Agenzia per l'Italia digitale, a public agency established in Italy by the Monti government (2011–2013). It pursues the highest level of technological innovation in the organization and development of PA.

2 ENEA. Main Italian National Agency for New Technologies, Energy and Sustainable Economic Development. It promotes important research on these issues.

laboration between the main actors of urban processes, creating a roadmap capable of providing the necessary tools to outline a "shared" profile of the smart city. See also the debates on Forum PA [6] , in Italy,[3] and the related events, aimed at the implementation of digitization of the PA.

10.2 The question: urban planning between technocracy and governance (literature review)

The connection between governance and data collection and management would be significantly shared among the various actors because one of the central points seems to come from the strategic use of the large and growing amount of data that new technologies allow us to acquire and use in various ways and for different goals. It is difficult to ignore that the issue of governance of "the whole process," as decisive "frame" for an urban development that wants to remain within truly "democratic" scenarios, is directly linked to these two aspects, i.e., data "acquisition and management" (control and use of big data is one of the most relevant issues for the twenty-first century).

If the smart city is an urban structure where the most innovative information technologies and digitization can accelerate the growth of urban welfare and improve the overall quality of life of its inhabitants through controlling and using an impressive amount of data and, consequently, through evaluating social phenomena in a more proper way, then it is easy to understand that it is right and dutiful to observe and consider how relations between citizens and rulers must change, and how this can be done.

In fact, the capacities of government increase, and for this reason, the citizens do not seem to effectively public and private management of the enormous quantities of data, on which basis they propose (or rather would seem to propose) political strategies of social and economic development, as well as land infrastructure decisions.

In essence, all this could change the characteristics of "democracy" which, at least in the West, is still considered the most acceptable form of government.

Parag Khanna's text[4] Technocracy in America: Rise of the Info-State, seems to address this problem. It is also strange that the title has been translated into Italian as "Rebirth of City-States," which does not seem to clearly reflect its underlying aims.

---

3 Forum PA is a company of the Digital 360 Group. It has been working for three decades to stimulate the digital growth of the PA, encouraging the meeting between PA, companies, researchers, citizens.

4 C. Ratti, an internationally renowned engineer and architect, teaches at the Massachusetts Institute of Technology, where he founded and directs the Senseable City Lab. He is the owner of Studio Carlo Ratti Associati, whose Italian office is in Turin.

Actually, in the text, technocracy is a key term that allows one to immediately understand the relevance of this topic for city government. However, the author mainly refers to the situation of the US PA.

While analyzing various forms of government (direct or indirect representative democracy, ideological leadership, dictatorships, technocracy, meritocracy, etc.), the author claims that, unlike all others, a "technocratic" government is rooted in expert analysis and long-term planning rather than in improvisations without perspective which are typical of populism and certain "unconsolidated" forms of democracy. He emphasizes the prevalence of such forms of government where the *criterion of representativeness* (seen as the harbinger of continuous and wearying disputes on ideological statements) takes precedence over *those modalities of public administration which are more focused on providing the services* the citizens need, on the basis of reliable data and with ever greater immediacy.

One of Khanna's assumptions is that "numbers and democracy" should be given the same weight. Therefore, governments at any level should aim to effectively meet people's needs, using medium- to long-term scenarios and ultimately bringing together *democratic inclusiveness and management efficiency*, characterized by a greater emphasis on "technocratic" profiles.

In support of these claims, the author cites Switzerland ("direct" European democracy, bottom-up) and Singapore (Asian technocracy, top down), stating that the two states have the best values among coefficients' international statistics to assess the well-being level of citizens. Both democracy and knowledge can be put together, when knowledge implies that the rulers have historical, current, renewable, verifiable "data." The smart city (and even more digital city) debate and experiences seem to indicate that this model of governance is possible.

If this topic were to be extended from smart cities to national models of government and in particular to Italy, you could find, for example, that, in times of strong social hardship, the leadership of national governments in Italy was very often entrusted to nonpolitical personalities that had unquestioned technical competence, especially in the economic field. They were authentic technocrats coming from eminent schools of economy and, often, former governors of the Bank of Italy well known in Europe.

While describing the economic and political state models from an evolutionary standpoint, Parag Khanna asserts that nowadays social and economic development is mainly focused on large metropolitan areas. He suggests that the best performances, technologically advanced in management, are found in what he defines as "info-states" (postmodern democracies).

Advanced technological facilities would allow it to achieve better results toward the market. The two mentioned states often have a much more direct, almost real-time relationship with their citizens (referendums, petitions, investigations, public workshops, etc.).

The fact is that ideological leadership has been abandoned in favor of strong pragmatism. The communities must seriously assess whether they give priority to various forms of exasperated democracy (but with an objective lack of welfare) or to forms of government that are able to effectively meet their needs but, at the same time, have substantial democratic profiles.

This discourse could not overlook Italy, where widespread "populism" has gained much of the credit previously entrusted to the traditional "political class." Khanna's [8] remarks make us wonder if democracy can guarantee the success of a country or if the growing integration of *technocratic positions* into the roles of governance can lay the ground for a new and more fruitful link between administrators and citizens.

These steps, Khanna suggests, could take place if we began to deepen the understanding of such pairs as politics and governance, democracy and services, process and results in order to understand how it is possible to get the best from each component, selecting the sense of "process" from democracy and the sense of "result" from technocracy.

But the most interesting aspect of this analysis is that neither of the two main options should prevail at the expense of the other. If they did, they would delegitimize themselves.

The author does not hide that, in his opinion, "in the long run, the quality of governance matters more than regime type" M. P. R. Bolívar. 2017 [4] (rather questionable consideration), and argues that: "Too much politics corrupts democracy; and too much democracy gets in the way of policy. Politics is about positions, policy about decisions; democracies produce compromises, technocracies produce solutions; democracy suffices, technocracy optimizes."

Still with regard to the current situation in Italy (first semester of 2021), it seems possible to make an immediate reference to the structure of government that the president of the republic proposed to the country: the government was entrusted to an economist of world renown, a decidedly pro-European former President of the ECB, a true technocrat with a staff of technical, experienced and nonpolitical ministers. He makes decisions on strategic issues while referring minor choices and more explicit political interventions to the other members of the council in the name of partisan representation of the electorate.

In support of the above-mentioned statement, P. Khanna points to the remarkable social and economic results achieved by Switzerland and Singapore and considers them as the two best prepared forms (democracy and technocracy) to provide excellent forms of welfare and development for citizens. These types of government use scenarios and combine forms of direct democracy (local governance situations, to which immediate responses are to be given): from government with consent to government with consultation (crowdsourcing).

Therefore, democracy must be characterized by a very different political dialectic compared to the current state (explicit reference to some European democracies, first and foremost Italy), which starts from recognized and formally shared consistent analytical bases and provides adequate and timely forms of knowledge, representing the quantity and quality of the phenomena in a certified manner.

These grounds can be the instrument that allows democracy not to succumb to constant forms of increased participation by citizens, which often degenerate into very rigid contrasts and cause repeated suspensions of procedures and variants of products.

Actually, from the postwar period onward, in the so-called Western democracies, but not exclusively, the phenomenon of urban governance has always been accompanied by numerous analyses and collections of data on social and territorial realities, which have been poorly utilized in practice. The reasons were varied: general unpreparedness of the ruling classes to change their decision-making behavior and their tendency to rely mainly on traditional applications, which guaranteed more fruitful mediations between political parties; low theoretical and technical training of the specialists, so that the so-called case studies were never used as stimuli to create more innovative tools. As a result, the new policies to be implemented had to be systematically supported by further analysis and research that often led to different results and gave rise to a system of continuous uncertainties in governance.

While discussing the substantial fragility of governance based predominantly on formal democratic structures, P. Khanna notes that governance, which uses innovative technologies for data collection and analysis and is entrusted (one of the central points of this discussion is the "level" of this reliance) to technocratic experts, could represent a significant step toward achieving the "concrete" objectives that any form of public government aims for, namely the well-being of citizens. We will see, moreover, that in 2021 the author has partially integrated or modified this position due to global emergency situations.

Ultimately, a new type of governance is proposed. It is characterized by appropriate profiles of technocrat administrators or, better yet, by a decision-making system that uses a "close relationship" between political bodies – representatives and technocratic bodies. The latter provide qualitative and quantitative assessments for the proposed development scenarios. The general objective, therefore, is to launch a "technocratic" evolution of the political system, which is a very sensitive topic from a political point of view.

In fact, the article [10] discussed if the political system gradually acquired increasing contributions from the technocratic system, it would give rise to "utilitarian and meritocratic" goals, highlighting to what extent current political decisions are based on ideological contrasts and on poor knowledge of urban phenomena.

The utilitarian and meritocratic system of technocracy could thus offer a more rigorous depth of action to current forms of politics. In other words, political and administrative action would be more effective.

That is why, at the present moment, urban and territorial planning, marked by long-term procedures that inevitably deal with situations requiring more than partial revisions, demands a broad reflection on "who and how can manage" the databases in order to use them in a manner consistent with social demand.

But if a more rigorous scientific and technological profile of administrators is postulated, it is equally clear that technological innovations steadily expand in terms of the collection, storage and use of data. This creates a problem for the end users of welfare services, who should also be able to manage both outgoing and incoming data in ways that ensure the appropriate and desired use of those services by the communities.

10.3 Results and analysis of results: questions for the future

In conclusion, we wonder if the suggestion of P. Khanna is acceptable: "Direct technocracy is the best form for twenty-first century governance, because it combines an executive and a Swiss-style multi-party parliament with the Singapore public administration, led by utilitarianism and data analysis. A blend of democracy and technocracy, in other words assisted by technology."

The authors of this chapter aim to use some of Khanna's suggestions to develop the topic of governance, fully involved in a democratic social structure and giving real space to "participation," which can also be conflictual but still improves the physical and social well-being of citizens (not only users) through a qualitative and quantitative knowledge of phenomena that can activate both "vertical" (top-down) and "horizontal" (as well as bottom-up with multiple subjects) decision-making processes. Thanks to this strategy and the maximum use of new innovative ICT technologies, governance can, for example, redesign cities in time and space and in shared ways.

In the scenario [8] determined by an over-decade-long debate on smart cities, we can surely agree on some issues, such as the fact that we are building a "gigantic collective memory" on a global scale. The importance of this fact for other issues (taxation, democracy, political classes, meritocracy, etc.) is indisputable; we can see it in the media every day.

Updating previous research, the authors argue that, especially after the effects of the global pandemic, it is difficult to imagine the new "future city" without defining the political obligations of the state concerning governance. Following the idea of some scholars [2],[5] it is necessary to "explore the future" by designing the present, hypothesizing its eventual scenarios, inspecting them and admitting possible or probable transformations (the "futurecraft" of C. Ratti) instead of giving space to unreliable scenarios and visions.

5 Parag Khanna(2008) The Second World: Empires and Influence in the New Global Order Hardcover – March 4, 2008 by Random House, 1st Edition, ISBN -13: 978-1400065080.

We are living in a momentous time marked by the great transformation of communication systems between people and communities. We are transitioning from the first "global village" to a village where communication has acquired a more democratic character (though we also see its limits) due to both "vertical" top-down and increasingly "bottom-up" and "horizontal" pushes, as evidenced by the Internet and cyberspace. This space of "flows" is new compared to what existed just 50 years ago. It is interactive between the world of digitization, computer science and the physical world. This immediately raises the question of the effect it may have on the physical form of the city [6]. The spread of communication between subjects (not only people but also things) has completely changed the concept of physical "distance," even if some consider this a "fictitious" argument. In fact, it is necessary to understand how much the Internet makes it unimportant to define the location of interconnected subjects and things (e.g., smart working, remote fact-checking and social meetings). On the contrary, we see the continuous growth of cities, even if it is not generalized and is more prevalent in what was once called the Third World. This continuous urban growth does not convince us that the most current forms of communication cannot make the urban sprawl phenomenon possible or real.

The human being is a "collective" animal who needs a real, physical community to live. The contact and life in the city do not deny the opportunities of cyberspace. It is the "triumph of atoms and bits" discussed by C. Ratti (2017) [9]. Thus, the urban space is a set of "virtual and physical places"; networks and places combine and give rise to changing urban "forms" that are different on a physical and functional level. This is the smart city in necessary combination with the digital city, where the Internet allows a different use of the city than the one imagined 50 years ago. People and things are, in fact, real-time "sensors" of the evolution of urban reality (ubiquitous computing) or at least some/many of its fundamental components. These include "public and private spaces." Therefore, from the analog to the digital era, the container has taken on forms that are not only physical but "different in time, even short, and flexible in functions." The concept of city "form" changes, acquiring a more complex and variable meaning over time [7]. Thus, the city evolves/mutates through *data-driven* processes, and the creation and use of (digital and physical) urban space may be (in terms of time impossible to predict) coordinated/unified by one or several computer platforms that immediately raise [9] fundamental questions. Is civil hacking possible or likely (to "hack" the city, according to S. Sassen)? What kind of governance is desirable and acceptable for a very complex process driven by knowledge and new technologies? Here it is easy to imagine a programmed and changeable architecture, which includes a "sentient space" (C. Ratti) [11]. The exponential growth of big data offers a glimpse of enormous research spaces to understand the effective functioning of cities, the "consistency/resilience" of their forms and the "quality" (political-social coherence) of governance of these same data and of the entire data-driven process.

10.4 Conclusions

This whole system is based on an "intensive" use of technological instrumentation and its implementation processes. However, the worldwide phenomenon of the Covid-19 pandemic (2020–2021), along with the acceleration of climate change and significant migratory phenomena, introduced further elements that may become structural "components" of more appropriate considerations on the changing relationship between decision-makers and citizens.

This seems to emerge [9] from P. Khanna's most recent contribution on the theme of world evolution: "MOVE" (4) with the significant subtitle "The Forces That Are Uprooting Us and Will Shape Humanity's Destiny" (2021).

The authors of this chapter point out that COVID-19 pandemics, unstable economies and climate change have first troubled and then interrupted the international flows of goods and people. The map of the world is being redefined, and we must understand what we are escaping from, why and where we want to go. In this sense, smart cities with their system of potentially good relations between decision-makers and citizens can provide positive answers, provided that they offer the opportunities required by future citizens, i.e., those whom P. Khanna [10] calls "digital natives." In his analysis, he identifies some phenomena that were already in place before the current pandemics (smart working, coworking, coliving and home sharing), especially child-friendly cities that appear to be the first positive answer to questions that concern young people all around the world. He defines these cities as "community cities," where at least embryos of a Bill of Rights are recognized. First among the latter are employment, security, including health security, mobility and limits to the ubiquity of technologies. All these can place democracy and technocracy on an inclined plane, giving advantage to the latter.

So, smart living is "the only way out to hopeless panoramas made of decrepit infrastructures, overcrowded blocks, raging traffic and rampant corruption."

With regard to this contribution, we are eager to understand the current status of big data governance and its probable evolution at a time when data-driven approaches, based also, if not above all, on the bottom-up and horizontal models, become possible. In particular, our interest is in "how" the public space can be programmed and put into practice through a form of governance that increasingly takes this model into account.

In short, we want to stress that the data-driven process, which has been set in motion, needs to evaluate in parallel the validity of the current forms of governance (of the process) so that these forms do not deny from the beginning the main point of the problem, i.e., the necessary form of substantial democracy of the entire operation. Without this necessary democratic profile, this can grow into the negation of the smart city (see P. Khanna, 2021) for which the meaning of cities with increasing forms of urban efficiency can no longer clarify the deepest sense of this phenomenon. On this issue some analysts [3] claim: "Digital transformation as ecosystem . . . is crucial.

If not governed, it becomes the harbinger of . . . terrible discrimination . . . of great violation of rights . . . devastating monopolies."

This is a crucial issue, especially in Italy. We should seriously ask whether our current political and administrative system is capable of providing a positive response to these concerns.

The scenario in Italy is changing. Some cities have become aware of the phenomenon and its dangerous aspects and want to protect themselves by experimenting with virtuous processes. They understand that the multiple technical definitions, smart, intelligent, digital cities, etc., are of little use if we do not appreciate the relationship these processes may have with the concept of democracy [6].

In this study, the multiple issues analyzed by P. Khanna immediately acquire prime importance. Some of these issues include the current relations between the forms of politics, administration and their relationship with the citizens.

The most preferable hypothesis suggests incorporating into the realms of politics and administration, which are still chosen through traditional methods, other forms of selection that favor meritocracy and, consequently, "official" forms of technocracy.

Options derived from factual evidence, primarily the consequences of the pandemic and climate change phenomena, characterize the current forms of urban and territorial management. These facts can be substantiated by the term "smart city," where, however, it should be applied to a content that concerns not only the efficiency of the current settlement model but also the management, precisely in view of the complex process that can be seen today in the big data collection, procession and use. In this process many actors are involved, including the continuous strengthening of the category of citizens as well as users. And, since such transformations require decades of work, analyses must mainly concern the future users (digital natives) of smart cities who will require its adaptation to their basic needs of life.

A governance model [4] must be fully understood and applied both in the legal setting and in the definition of a city with a high quality of life and, at the same time, on a human scale (welfare).

It seems that many local governments have come to realize that the "democratic" side of this process is not secondary at all to making the city more efficient, given that "efficiency" is not an isolated matter but is related to goals shared with the "citizens," that is, with the entire community. In fact, without denying the validity of an "efficient" city for users only, the distinction between them and "citizens" is the whole essence of a "smart" city [5].

References

[1] C. Bellone, P. Ranucci and V. Geropanta. The 'Governance' for Smart City Strategies and TerritorialPlanning. In: P. Vasant, I. Zelinka and G.-W. Weber Editors. Intelligent Computing & Optimization.Advances in intelligent systems and computing, p. 76–86, vol 866, Springer, 76–86, 2018. Advancesin Intelligent Systems and Computing, Springer, Cham. https://doi.org/10.1007/978-3-030-00979-3_8

[2] Parag Khanna(2008) The Second World: Empires and Influence in the New Global Order by Random House, 1st Edition, ISBN-13: 978-1400065080

[3] Parag Khanna(2021) Move: The Forces Uprooting Us–by Scribner, First Edition, ISBN-13: 978-1982168971

[4] M. P. R. Bolívar. Governance models for the delivery of public services through the web 2.0technologies: A political view in large Spanish municipalities. Social Science Computer Review, 35,2017, https://doi.org/10.1177/0894439315609919

[5] N. Komninos. The Age of Intelligent Cities: Smart Environments and Innovation-for-All Strategies. Routledge, London & New York, 2014, DOI:10.4324/9781315769349

[6] F. Naselli. Nuove Forme Della Qualità Urbana, FrancoAngeli, Milano, 2012, ISBN-13: 978-8820414870.

[7] N. Komninos. The Age of Intelligent Cities: Smart Environments and Innovation-for-All Strategies. Routledge, London & New York, 2014, DOI:10.4324/9781315769349

[8] P. Khanna. La Rinascita Delle Città-stato Come Governare Il Mondo Al Tempo Della Devolution, Fazi (Ed.), 2017, ISBN-13: 978-8893252294

[9] C. Ratti and M. Claudel. La città di domani Come le reti stanno cambiando il futuro urbano. Passaggi Einaudi, 2017, ISBN 9788858426265

[10] A. Caragliu and P. Nijkamp. The impact of regional absorptive capacity on spatial knowledgespillovers. Tinbergen Institute Discussion Paper No. 08–119/3, Tinbergen Institute, Amsterdam,2008, Retrieved from: https://www.econstor.eu/handle/10419/86816

Ekin Can Erkuş* and Vilda Purutçuoğlu

Chapter 11
Outlier detection in biomedical data: ECG-focused approaches

Abstract: Outliers are among the most important data behaviors in time series data and have exceptional importance in biomedical data analyses. They may either represent an unwanted problematic aspect or a core data behavior. Detecting and correcting the problematic samples in data may improve further analyses such as model performance and diagnostic accuracy. On the other hand, locating the outlying samples in the main data behavior may help with pattern recognition in the data. Therefore, the outlier detection step in the preliminary analysis holds a critical key to the rest of the analyses. This chapter investigates the three types of outliers found in the biomedical data by providing examples from the electrocardiography data modality.

Keywords: Outlier detection, ECG Signalling data, R peaks

11.1 Introduction

Biomedical data modalities often have distinctive behavioral patterns, such as a PQRST structure for electrocardiography (ECG) data. Detecting such structures is the key to revealing the data modality and type [1]. Moreover, distinctive structures inherit critical features for various data conditions such as disorders, event-related responses or environmental changes. Hereby, the detection of irregular conditions mainly depends on identifying the unique structures in the data [2]. The literature on the PQRST structure detection in ECG data has many methods, and most of them primarily focus on detecting a single structural component [3–6]. The most common approaches are based on locating the R peaks and continuing with the whole structure's identification based on the detected R peaks [7]. Here, both the initial location and the identification parts require some algorithmic steps to be implemented, whereas several components in the raw data may reduce the efficiencies of these steps, such as noises, artifacts or event markers, which constitute a variety of outlier types. Outliers or anomalies in data mining terminology, as a definition, refers to the data samples that do not behave like the rest of the data samples in terms of statistical properties, distribution, periodicity or

***Corresponding author: Ekin Can Erkuş**, Huawei Turkey R&D Center, Intelligent Applications Department DC, Turkey, e-mail: ekincanerkus@hotmail.com. https://orcid.org/0000-0002-2445-5929
Vilda Purutçuoğlu, Middle East Technical University, Department of Statistics, 06800, Ankara, Turkey, e-mail: vpurutcu@metu.edu.tr https://orcid.org/0000-0002-3913-9005

https://doi.org/10.1515/9783110717655-011

other factors with unexpected behavior [8]. Hence, depending on their intensities, they may change the statistical properties of the data, may distort the data mapping in the frequency domain or may behave as a faulty component in the raw data [9].

Classification-based machine learning algorithms, data modeling and diagnostic analyses are very susceptible to the presence of outliers [10]. Accordingly, the detection and the removal or the correction of such unwanted constituents in the raw data play a crucial role in the rest of the analyses, and outlier detection algorithms used for this purpose are performed as part of the preprocessing stage in biomedical data analyses. Due to their underlying importance, the effects of detecting the outliers and their corrections from the overall analyses have been intensively studied in the literature [11–14] via machine learning approaches [15, 16], modeling-based approaches [17, 18], and various other methods for data analysis [19]. Figures 11.1 and 11.2 present a basic example of an ECG dataset, illustrating how a single outlier can affect the overall data mean by comparing the overall mean values of the same data with outlier outlier-corrected forms, respectively. Note that the mean of the data in Figure 11.1, i.e., the mean without outliers, is found as −8.59, while the mean of the data in Figure 11.2, i.e., the mean with a single high-amplitude outlier, is 94.43. This difference increases when the amplitude of the outlier increases or the sampling rate of the data decreases. As a result of this, by changing the statistical properties of the data, the outliers reduce the effectiveness of further analyses. Moreover, by acting as components in the data, outliers also reduce the performance of feature extraction algorithms.

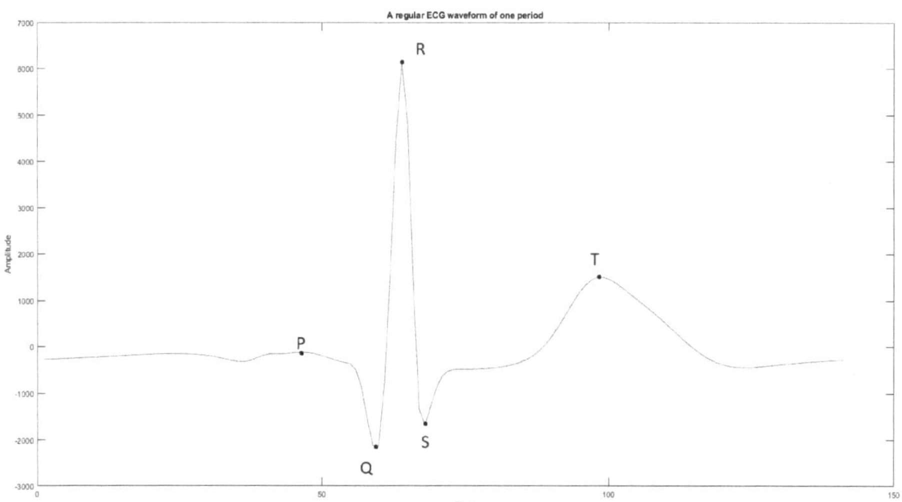

Figure 11.1: A regular PQRST structure of ECG data.

The outlier detection accuracy may change with the data type, data length, number of outliers and outlier behavior [20]. Such outlier behavior can be defined under several

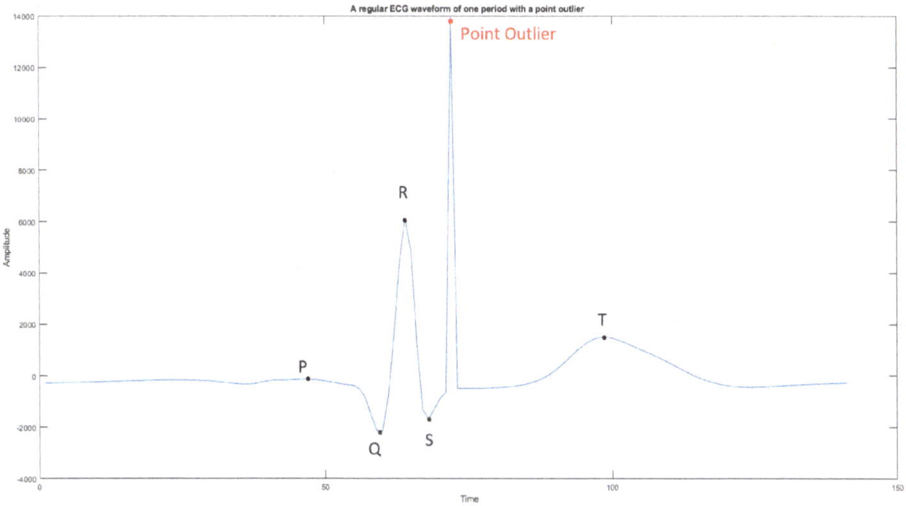

Figure 11.2: A regular PQRST structure of ECG data with an outlier point added.

categories depending on their characteristics, such as periodic, single or clustered, and trended structures [21]. Moreover, outlier detection algorithms specialize in specific types of outliers in various biomedical data. Therefore, the selection of the optimal outlier detection algorithm does not solely depend on outlier behavior but is also directly affected by the data modality and type. Hence, the types of outlier detection methods and the applicable biomedical data must be investigated before the analyses [22]. A brief section about the R and QRS detection algorithms is given in the next part. Then, the following section categorizes the types of outlier detection methods and briefly presents their characteristic features. Finally, the last section before the conclusion applies those outlier detection methods to the proper biomedical data modalities, particularly for the ECG data.

11.2 A brief knowledge about R peak and QRS detection algorithms

A normal ECG rhythm has the distinctive individual components of the PQRST structure, where the proper appearance of each component is a requirement for a healthy heart. Although there are many other defined components in the ECG data, the components P, Q, R, S and T are the most commonly used and analyzed constituents of the ECG data modality. Further considering the distinctive statistical and signal-wise properties, the QRS structure or the R peaks have the most popular features to reveal the behavior of the ECG data [23]. As a result, most ECG studies often focus on detecting either R peaks or QRS structures before the advanced analyses. The characteristic behavior of a PQRST structure can be found in Figure 11.1, where each component is identified and marked on the data.

The QRS structure refers to the ventricular depolarization in a healthy cardiac measurement. Several disorders are related to distortions in the QRS structure in ECG data, such as hypertrophy, hyperkalemia or infarction [24, 25]. Some algorithms have been proposed to detect the QRS structure to diagnose cardiac disorders efficiently [26]. The detection of the QRS structure is often performed as a posterior to the R peak detection and involves most of the steps to reveal the R peak locations. As the major component of the QRS structure, the R peak is distinctive due to a larger fold change in data mean and hence can be considered the main marker of a heartbeat period [27]. The distance between adjacent R peaks corresponds to a cardiac cycle, and the frequency of the R peaks in an ECG data modality is named the heart rate. The change in the heart rate per beat is named the heart rate variability (HRV). Hence, the deterioration of HRV over time can be considered the main indicator of rhythm-related disorders, which makes the accurate detection of R peaks an important part of ECG data analysis [28]. There are many algorithms for R peak detection in the literature, which can be categorized as:

- statistical approaches,
- derivation-based approaches,
- window-based approaches,
- frequency domain or transformation-based approaches,
- learning-based approaches and
- combination of these [29–31].

Statistical R peak detection approaches mainly focus on the high mean difference of the R peaks in the ECG data [32]. After preprocessing steps such as detrending, some parametric approaches, such as standardized scores, or nonparametric approaches such as box-plot algorithms can be used. The key idea is to determine a threshold value and consider the samples above that threshold value as the R peaks in a detrended ECG dataset. Statistical methods are often combined with derivative-based approaches to obtain better R peak detection rates. The samples above the threshold value are further investigated for their first derivative values. A simple drawback of derivative-based approaches is noise, which disables the monotonic increase or decrease in the data. Therefore, before applying the first derivative, noise elimination or smoothing operations are performed. The Pan-Tompkins algorithm is one of the most commonly used derivative-based approaches and involves a filtering operation [33].

Moreover, some algorithms combine both the first and second derivatives for even higher success in the R peak detection [34]. On the other hand, frequency domain or transformation-based approaches often provide more robust results than statistical methods in the R peak detection. A popular transformation-based approach for R peak detection uses the Hilbert transform, where the data are transformed into the Hilbert domain or an altered frequency domain. The Hilbert transform approaches are sometimes used after the differentiation operation [35] and sometimes with an adaptive matched filtering technique [36]. Yet another frequency domain-based approach, the wavelet transform, is often applied to estimate the periodicity of the ECG signal by using transformations of the frequency domain in sliding windows through the data [37]. The wavelet transform is a useful R peak detection algorithm for ECG data, especially with low HRV values [38]. In most studies, the wavelet transform is used after the digital filtering process. The cut-off frequencies of the filters and the filter coefficients should be well arranged to avoid losing any valuable frequency components. This operation may be useful in the following feature extraction step, such as spectral features [39].

Recently, some new hybrid R peak detection algorithms have been developed based on combinations of the aforementioned approaches with a few modifications [40]. Most of them aim to improve the accuracy in detecting R peaks, although they have a drawback in computational times. On the other hand, some focus on improving the computational times by either simplifying the original algorithm or optimizing the algorithm steps. One of the recently proposed R peak detection algorithms is based on the hierarchical clustering of the amplitude and slope features of the samples [41]. With this method, the possible R-wave clusters estimate the R peak locations by further analyzing the slopes within the R-wave clusters. Their results show that the accuracy in detecting all of the R peaks in the MIT-BIH arrhythmia database is higher than the wavelet transform and the moving average filter-based approaches. However, due to both learning and signal processing steps, this method seems computationally complex, which may prevent it from being used as a real-time R peak detection algorithm.

Another recent method utilizes the Hilbert envelope function to obtain the locations of the R peaks in the Hilbert space [42]. First, they apply a band-pass filter to eliminate some noises and variations in the data. Then, by applying the Hilbert envelope function, the peaks appear at the expected locations of the R peaks. Hence, the R peaks are detected by applying a threshold and separation rule to those peaks. They report that the accuracy of the R peak detection is high, and the algorithm is simple to use. However, selecting filter components requires preliminary work before the analysis, and eliminating the low-frequency components may eradicate some valuable information in the data.

Another method, the frequency domain-based outlier detection (FOD), was originally developed to find periodic outliers in the time series data and is used as the R peak detection algorithm for simulated ECG data with zero HRV. They report that the algorithm is almost as fast as some parametric algorithms with perfect detection of periodic outliers in the time series data. However, FOD uses the fast Fourier transform (FFT) for the entire data; hence, variations in the heart rate cannot be detected. A possible modification for the FOD algorithm could be the implementation of the short-time Fourier transform instead of FFT. Another possible solution for the FOD algorithm is to apply the FFT in moving windows, labeling the possible peaks and evaluating the R peak locations on the overall data. However, this may increase the computational complexity of the modified FOD while enabling the handling of HRV.

11.3 Most commonly used outlier detection approaches

There are many types of biomedical data modalities and data behaviors; also, there are many outlier detection algorithms. Apart from the variety of outlier detection algorithms, several classifications of such algorithms are made. Firstly, depending on the labels of the outliers to be detected, the algorithms can be categorized into supervised, semisupervised and unsupervised methods [44]. This kind of classification is mostly used for learning-based approaches. Another popular classification of outlier algorithms can be made depending on the number of data channels in which the algorithm operates. Hence, outlier detection algorithms can also be classified as univariate, bivariate and multivariate [44].

Moreover, depending on the statistical properties of the outlier detection algorithms, they are grouped into two main classes: parametric and nonparametric methods. Apart from these main classes, some hybrid or irregular types of outlier detection algorithms are built for specific data types [45]. The summary of the classification of the outlier detection methods is provided in Table 11.1.

Table 11.1: Different categories of outlier detection methods according to their classification criteria.

Criteria	Methods
Labels	Supervised, semisupervised and unsupervised methods
Number of channels	Univariate, bivariate and multivariate methods
Statistical properties	Parametric and nonparametric, hybrid or irregular methods

The parametric outlier detection methods form the main basis of outlier detection algorithms and are generally fast with data-dependent behaviors. These methods rely on prior knowledge of the data distribution, where some parameters can define the distribution. For example, the well-known Gaussian distribution has mean and variance parameters, while the Poisson distribution only has the rate parameter [46]. Each parameter that forms a probabilistic data distribution has a major effect on the sample behavior. In contrast, the parameters are mostly dependent on the nature of the data for real biomedical data modalities. Specifically, for time series, most parametric approaches rely on the stationarity property of the data, where the statistical moments of the data remain stable for predefined time frames or under certain circumstances [47]. Most biomedical data are nonstationary signals as biological processes have stochastic behavior, and the parameters generally change over time. For example, in the ECG data modality, most of the time represents the nonstationary behavior due to changes in the heart rate over time and deviations in the independent nonlinear structures.

On the other hand, only some small frames of the ECG signals may show wide-sense stationarity, such as the heart rate and the individual structural behavior remain stable within the windows, which is not attainable in real data measurements [48]. In biomedical data analyses, the most commonly used parametric outlier detection algorithm can be considered the *z*-score, which has a fast implementation and is often used as the standardized value or under some custom definitions without crediting the *z*-score algorithm [49]. Here, under the Gaussian distributional assumption of the data, the distribution's tails after some threshold are considered outliers. Most of the standardized value algorithms depend on the localization of each sample in the data, which can also be considered an outlier score of the sample [50]. By applying parametric approaches without preprocessing steps, the R peaks are also marked as outliers and the correctly marked type I outliers in the data. Hence, the algorithm can be modified by using several more steps to decline the R peaks by definition.

On the other hand, it can also be modified to detect only the R peaks as an R peak detection approach instead of an outlier detection method. Due to its low computational complexity, parametric approaches like standardized value methods can detect outliers with high amplitudes in nontrended and normalized biomedical data [51]. Moreover, standardized value methods can also aid other outlier detection methods in generating outlier scores or providing prior knowledge of outlying structures as a preliminary step in some more complex algorithms [52]. Instead of parametric outlier detection methods, nonparametric outlier detection methods are more suitable for ECG data analysis due to the nonstationary behavior of the data unless prior processing steps are applied [53].

The learning-based approaches often require an initial feature extraction step where the properties of the features are trained to generate a model or to construct a foundation for the features of the upcoming data. Feature extraction is a large area of research in which useful features are extracted depending on the data modality, data conditions, and the focus of the research [54]. In biomedical data applications, features are often used in multivariate forms to generate multidimensional models, which generally increase the accuracy of the classification. In ECG data analyses, data are often segmented into the main structures, such as PQRST, and each component is investigated for its features. After the features are obtained from those components, outlier detection analyses can be performed to find the data segments whose components behave abnormally [55].

Several subcategories of unsupervised outlier detection methods can be based on main algorithm frameworks, such as clustering-based, classification-based, histogram and kernel-based approaches. The classification-based methods aim to separate data samples into classes by maintaining optimal partitioning according to the cost or the expectation as a process of optimization. Some well-known classification-based outlier detection algorithms for biomedical data analysis are support vector machines (SVMs), perceptron, regression and tree classifiers. According to their values, the samples or features are divided into two or more groups to be labeled as outliers, nonoutliers or other subcategories. Selecting the features to be used in classification-based outlier detection algorithms is important to obtain better results. Hence, before applying the outlier detection algorithm, the most distinctive structures in the data are investigated, and the most relevant features that reflect the effects of the outliers are suggested for selection [56].

Clustering and histogram-based frequentist approaches form a large category of outlier detection methods, which can be grouped as the learning-based approaches and are mainly used in machine learning approaches [57]. Their working principles for outlier detection rely on forming groups of data samples based on the variety of their features. Hence, any sample that does not belong to any group under predefined conditions can be considered an outlier. The most commonly used clustering-based methods in biomedical data analyses include variations of k-nearest neighbors, k-means, k-medoids, expectation-maximization and dynamical clustering algorithms. As the formation of clusters or groups of features is crucial to detecting outlying structures in data, choosing the best parameters is an important challenge in clustering algorithms [58]. Therefore, such algorithms often apply an additional pre-step for the parameter selection, which naturally increases the complexity of the methods.

Histogram methods are nonparametric and are based on the frequency of the occurrence of outlier amplitudes. The main idea of such methods refers to the rarity of the outliers. The less likely the amplitude anomaly occurs, the more likely the samples with that amplitude are outliers. Hence, it is often suggested to use histogram-based methods for small numbers of outliers with high amplitude in the data [59]. The histogram method requires a set step size to discriminate the amplitude values from each other. Since the step size may drastically change the shape of the histogram plot, choosing the right step size is crucial for detecting outliers correctly [60]. Therefore, as the smaller step sizes often result in better resolution, larger step sizes can be used to reduce the complexity of the methods for data with preferably large sample sizes, albeit at the cost of resolution.

Kernel density estimation methods are similar to histogram methods, but kernel methods use kernel functions for each sample instead of representing the samples grouped in solid columns. The kernel functions are often two-tailed with a predefined mean and standard deviation. The sum of the discrete kernel functions results in the kernel density estimation. Similar to histogram methods, the amplitudes outside the threshold values are considered outliers [61].

Probabilistic methods can be considered a specific category within learning-based approaches, particularly supervised approaches [62]. Most probabilistic methods are based on the Bayes theorem, which depends on maximizing the probability of outliers being labeled [63]. The prior knowledge of the outliers in the data is of great importance since probabilistic models are either built based on the known behavior of the outliers or the overall data behavior. Hence, depending on the structure and flow of the data analysis algorithms, new data samples, intervals or data channels are tested based on the probabilistic models. The most common probabilistic approaches for biomedical data analysis are Markov chains and hidden Markov models [64].

The application of Markovian approaches to ECG data may require the investigation and segmentation of the individual waves in the data to create a probabilistic model of how the structures appear in the temporal order [65]. After the components are detected and cropped, a normalization operation is generally applied to such structures to align the values within some specified intervals. This step provides more robust learning for the model, eliminating most of the trending and level shifts in the related components. The model consists of clusters of normalized components and the probabilities of the temporal changes from one component to another, named transitions. The transition probabilities between each component are calculated using the transitions' occurrences in prior observations, where the most frequent occurrence in the prior observations refers to the highest transition probability in the Markov model. According to the model, the new data are tested, and the posterior probabilities of each component are found. If the posterior probabilities of the components fall below the threshold value, they are considered outliers [66]. Probabilistic methods often do not give deterministic results for the same set of data in each trial [67]. Therefore, the repetition of analyses under Monte Carlo experimentation generally provides the results to converge at some points. As the number of repetitions increases in the outlier detection case, the true positive rate for detecting the outliers is also expected to increase.

The rest of the outlier detection methods are often developed for very specific behaviors of the outlier types. With this, the accuracy in finding the outliers is generally higher than other methods, provided that they are used on the correct types of data and outlier behaviors. On the other hand, most such methods have multiple computational steps with higher computational complexity, making their computational times costlier than those of other basic algorithms.

11.4 Types of outliers and their effects

Depending on the distinctive behavior in which they dwell in the data, outliers are categorized into three types: type I, type II and type III outliers [68]. Apart from the behavior, another categorization is made based on whether the outliers have a single sample, i.e., point outliers, a sequence of continuous samples, i.e., sequential outliers, and periodic outliers in the data [69]. As the outlier types vary, the proper outlier detection methods to detect each type of outliers also vary. Hence, investigating the outlier type in the data should be the first step of the outlier detection process. Accordingly, in Table 11.2, a summary of the outlier types and the appropriate categories of outlier detection algorithms is presented.

Table 11.2: Types of outliers and preferable methods to detect them.

Type of outlier	Amplitude or periodicity characteristics	Suitable outlier detection categories
Type I	High amplitude in the overall data	Parametric, nonparametric and learning-based
Type II	High or irregular amplitude in the regular periodic behavior; normal amplitude in the overall data	Transformation or frequency domain-based and learning-based
Type III	Irregular, periodic behavior of the data	Transformation or frequency domain-based, learning-based and nonparametric

11.5 Type I outliers

Type I outliers are distinguished mostly by their different statistical properties or features in the data. Independent from the data modality, both point, sequential and periodic outliers can be counted as type I outliers. Most of the research fields consider the general definition of outliers as type I outliers. For instance, some engineering studies define outliers as samples that exceed certain thresholds. In contrast, some statistical studies define samples as outliers when the standardized values of the samples reach a pre-set quantile [70]. The most common appearances of type I outliers in one-dimensional biomedical data can originate from noise and motion artifacts, subject-related abnormal readings such as disorders, markers, instrumental defects and environmental changes during measurements [71–73].

On the other hand, in addition to the same outlier causes as in one-dimensional data, for two- and three-dimensional biomedical data, subject-related abnormal readings such as lesions, tumorous tissues or abrupt heat changes can be counted as outlier-creating phenomena [74, 75]. Specifically, motion or device-related artifacts are the most common appearances of type I outliers for the ECG data [76]. Both of these may significantly change the statistical values of the whole ECG data or its segments. Motion artifacts often last longer and may cause a trend or level shift in the data. In contrast, noise artifacts are distinguished from motion artifacts by their short-term appearances. They may prevent automated algorithms from detecting individual structures such as P, Q, R, S or T waves by mimicking one of them [77]. Figure 11.3 refers to a motion artifact of a type I outlier sequence in ECG data, and Figure 11.2 shows a point outlier of type I.

The statistically different behavior of outlier samples generally appears and is conspicuous as a difference in the mean value of the data. Compared to type II and type III outliers, type I outliers are relatively easier to detect by statistical outlier detection algo-

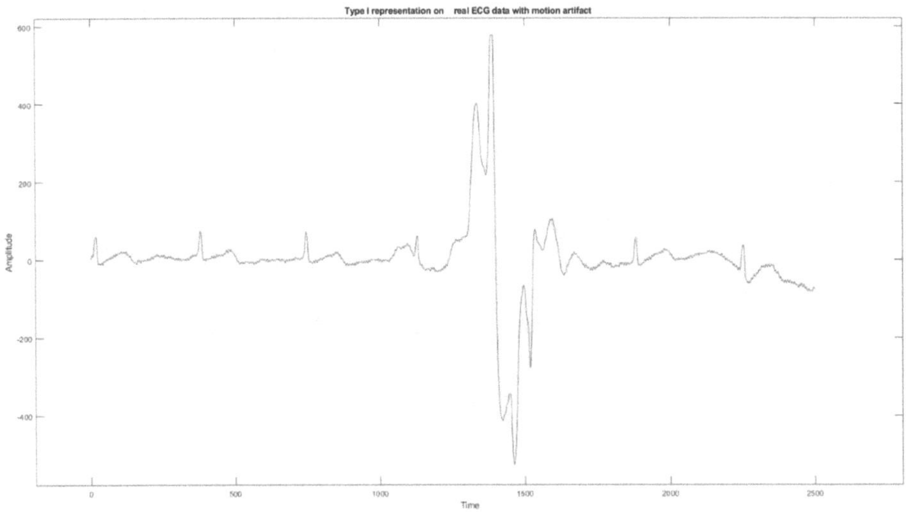

Figure 11.3: ECG data with a motion artifact in the form of type I outlier sequence.

rithms in the given dataset. Parametric outlier detection algorithms can be used after the basic preprocessing steps for the related data modality, such as noise elimination and detrending operations.

11.6 Type II outliers (contextual outliers and spatial outliers)

Type II outliers are often defined as the outliers with irregular behavior in a regular period of data behavior and are the most common outliers in ECG data. Data with type II outliers behave periodically with local peaks or troughs, and the outlying samples have irregular occurrences of such periodic behavior. Type II outliers can also be called contextual outliers. They can be suggested to determine by analyzing the temporal neighbors of the suspected data samples rather than examining the whole dataset. Most type II outliers can be considered inliers by the statistical definition of an outlier, where the outlying samples may not significantly differ from the overall data mean. Hence, standard statistical outlier detection approaches may not detect type II outliers unless several preliminary steps are applied. In ECG data, contextual outliers may appear similar to type I outliers in a single sample or sequential samples. However, the main difference is that the outlying samples have amplitude values within the normal data range instead of providing a global extremity point for the data. Therefore, type II outliers do not change the statistical properties of the data significantly but rather reduce the efficiency of the algorithms used on the data. Small-scale motion artifacts and device-related artifacts can be type II outliers. Although often not considered outliers, hum noise or data-related noises are the most commonly encountered type II outliers in ECG data. Here, such high-frequency noises can be considered periodic type II outliers. Although the artifacts have small scales, the noises may be found throughout the data. An example of a motion artifact-related type II outlier can be found in the data given in Figure 11.4, where the type II outlier sequence shows more irregular behavior than the rest of the PQRST structure of the ECG data and is colored red. On the other hand, noise-related type II outliers can be found in Figure 11.5, and the denoised form of the same ECG data can be found in Figure 11.6, but with a distortion in the R peak amplitude as a drawback of the denoising method.

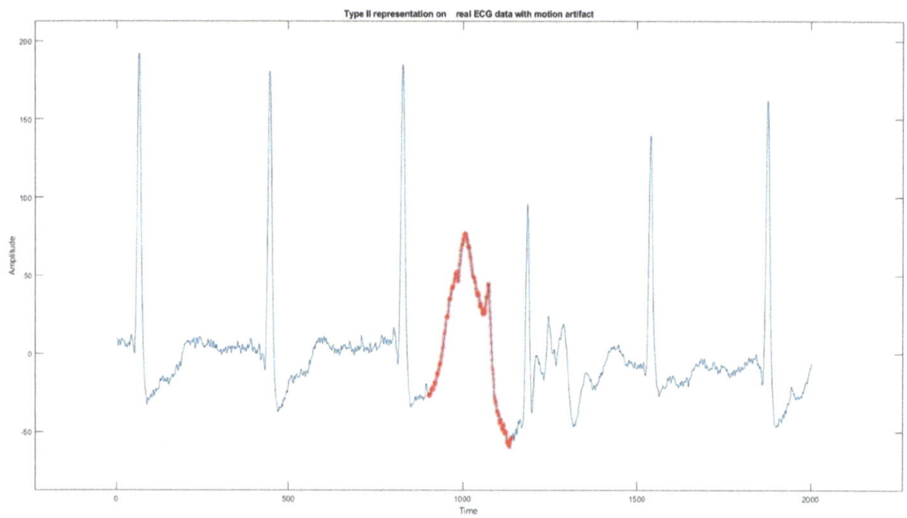

Figure 11.4: ECG data with a motion artifact in the form of a type II outlier sequence.

The detection of type II outliers requires more complex approaches than standard parametric methods. Firstly, either the period or the expected data behavior should be found using the frequency domain approaches, segmentation or moving/sliding window analyses. After the regular period of the data is found, further analyses take place in each period. Here, another direct solution might be the detrending of each period of the data segment and the application of periodic outlier detection methods in each period to locate the outlying samples [78].

Many methods in the literature can be used as type II outlier detection algorithms, even though they are not specifically designed for this purpose. For instance, modeling-based moving window approaches can provide robust results such as an autoregression-based algorithm ARIMA [79], a median-based nonparametric outlier detection algorithm such as the median-difference window subseries score [80] and an

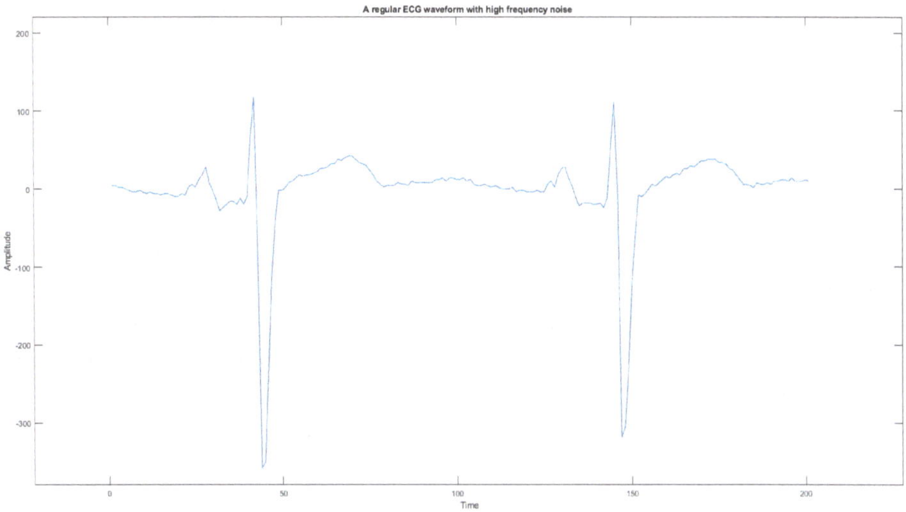

Figure 11.5: ECG data with noise can be considered a type II outlier.

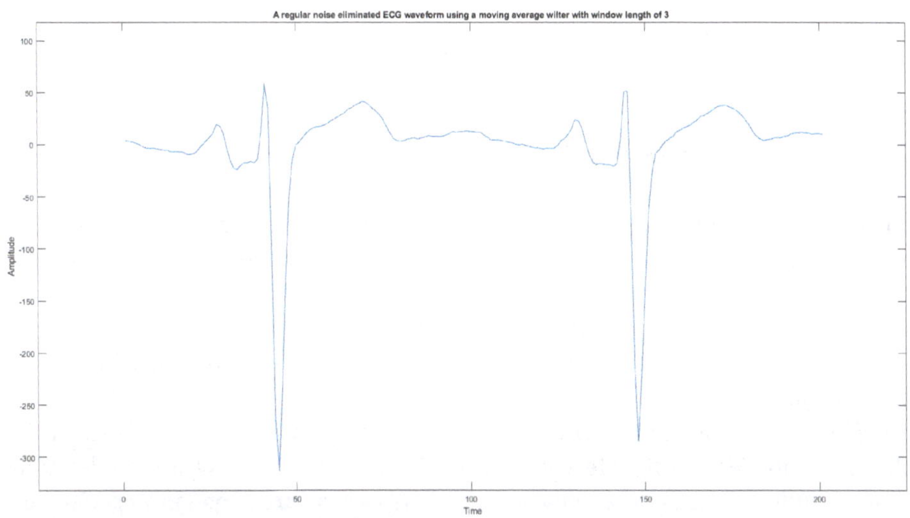

Figure 11.6: Denoised ECG data using a simple moving average filter.

algorithm that is based on a short/long-term pattern recognition [81] can be used to detect point type II outliers in periodic data.

Learning-based approaches can also be utilized in type II outlier detection, such as a deep learning approach on the encoder-decoder system [82], and a continuous learning approach called hierarchical temporal memory [83] has been used for ECG data.

Modeling and frequency domain-based approaches are also used to detect contextual outliers, such as the wavelet transform-based regression [84] and the FOD [85]. Moreover, combining the frequency domain approach with the autocorrelation structure, Fourier autocorrelation structure is highly efficient in detecting contextual outliers in multichannel ECG data [86].

Clustering-based approaches on ECG datasets with type II outliers generally perform the segmentation of heartbeats and group the similarly shaped beats together. A cross-correlation-based clustering algorithm, normalized cross-correlation clustering, presents an iterative approach with cross-correlation values and is iterated until a type II outlier is detected in a heartbeat cross-correlation where the heartbeat differs from the rest of the group [87]. Another popular clustering outlier detection algorithm is called the density-based spatial clustering of applications with noise, which is based on the sparsity of the abnormal segments of the ECG data [88].

Noise related to type II outliers can be removed by filtering the data and eliminating the data's selected frequency components. There are many applicable filters, but selecting the proper filter and its coefficients directly depends on the data type and the behavior to avoid losing some useful information from the data. Several studies are directly related to the filtering operation with contextual outliers, but many signal processing studies involve filtering operations on a large scale [89, 90].

11.7 Type III outliers (collective outliers)

Apart from the statistical properties, the periodicity of the outliers may be crucial to detect for seasonal or periodic datasets where the outlier behavior either appears periodically or is statistically different from the overall data. With this, some intervals in quasiperiodic data may occur, either representing the nonperiodic behavior or periodic behavior with different frequencies in a periodic dataset. As opposed to the detection and elimination process of outliers from biomedical datasets, such periodic behavior of the data generally represents the distinguishing characteristics of the data modality. Most of the literature about ECG data anomalies focuses on detecting type III outliers, known as anomalies under the cardiac arrhythmia categories [91]. Arrhythmia has a wide range of subcategories, and so do the causes, such as signaling malfunction, hormonal instability, nutrition dependence, cardiovascular system degeneration or disease-related malfunctions [92]. Each of the causes can be extended, but their main effect on the ECG data is to alter the regular periodic behavior of the ECG data. Moreover, such distortion of the regular periodic behavior of the data is sometimes caused by the temporal combination of the independent components of the ECG data, where the subintervals in the PQRST structure are often narrowed or prolonged in time, and the components constitute a new outlying structure [93]. A clinical example of an outlying periodicity from the rest of the periodic behavior of the data is a ventricular tachyarrhythmia interval during an ECG recording [94], where the heart rate increases along with a severe desynchronization of the PQRST structure, generating an outlying structure in the whole data, as shown in Figure 11.7.

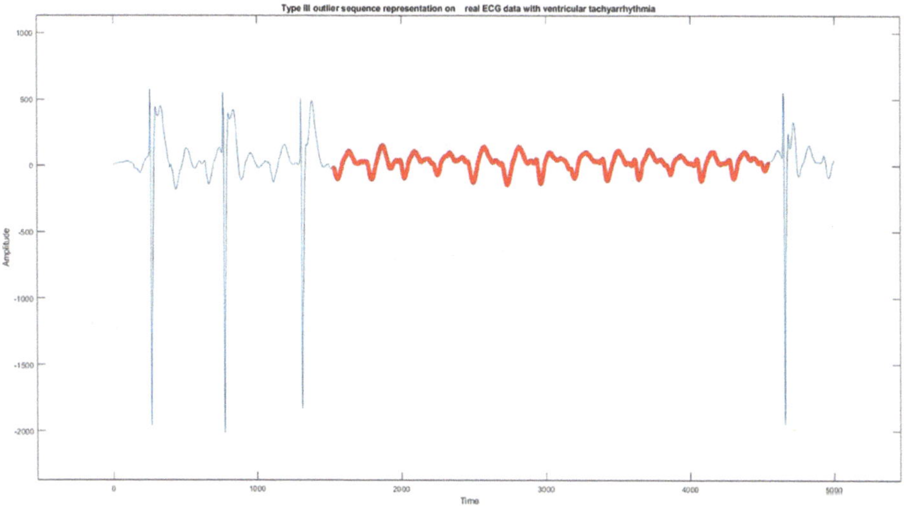

Figure 11.7: ECG data with ventricular tachyarrhythmia as a type III outlier sequence, which is marked in red.

Although each type of arrhythmia interval has characteristic data behavior, their detection processes in a regular heartbeat are relatively easier than detecting type II outliers. The most commonly used methods to detect such type III outliers in ECG data include frequency domain approaches such as the wavelet transform, variants of the Fourier transform or spectral analyses, such as moving window and autoregressive modeling approaches; learning and modeling approaches such as deep learning, neural networks and SVMs; clustering and density-related approaches such as k-means and k-medoids; distribution-related approaches such as the Kolmogorov-Smirnov test; and mathematical approaches such as derivative analysis [95]. Each of these approaches has a particular advantage in finding specific types of arrhythmia in ECG datasets. On the other hand, most of the R peak or the QRS detection algorithms can be used with some simple modifications, including the moving window or the frequency domain-based approach, to detect such anomaly intervals [96].

11.8 Conclusion

Outlier or anomaly has a very wide sort of definition. In other words, different research fields often use similar terms with different names, such as outlier types. From a statistical point of view, an outlier is a sample with a much larger amplitude than the rest of the data. On the other hand, two more definitions of outliers have been made, and they have been collected in this chapter. This chapter presents collective and brief information about the outliers encountered in biomedical data, where the examples and illustrations are based on the ECG data modality.

Outliers or anomalies reduce the stability of models and the efficiency of advanced methods implemented on the data. Hence, the outlier detection or the anomaly detection, and the elimination or the correction of any type of biomedical data are important preliminary steps for most further analyses. Each data modality has a specific behavior and different types of outliers. Therefore, selecting the most appropriate outlier detection algorithm is important in finding outliers while keeping the false positive rates as minimal as possible.

References

[1] J. Rodrigues, D. Belo and H. Gamboa. Noise detection on ECG based on agglomerative clustering of morphological features. Computers in Biology and Medicine, 2017. doi: 10.1016/j.compbiomed.2017.06.009.

[2] V. Chandola, A. Banerjee and V. Kumar. Anomaly detection: A survey. ACM Computing Surveys, 2009. doi: 10.1145/1541880.1541882.

[3] S. Katsigiannis and N. Ramzan. DREAMER: A database for emotion recognition through eeg and ecg signals from wireless low-cost off-the-shelf devices. IEEE-EMBS International Conference on Biomedical and Health Informatics, 2018. doi: 10.1109/JBHI.2017.2688239.

[4] M. Elgendi, S. Mahalingam, M. Jonkman and F. De Boer. A robust QRS complex detection algorithm using dynamic thresholds. In: Proceedings – International Symposium on Computer Science and Its Applications, CSA 2008, 2008, doi: 10.1109/CSA.2008.16.

[5] A. Illanes-Manriquez and Q. Zhang. An algorithm for robust detection of QRS onset and offset in ECG signals. Computers in Cardiology, 2008. doi: 10.1109/CIC.2008.4749177.

[6] R. Ruangsuwana, G. Velikic and M. Bocko. Methods to Extract Respiration Information from ECG Signals. In: ICASSP, IEEE International Conference on Acoustics, Speech and Signal Processing – Proceedings, 2010, doi: 10.1109/ICASSP.2010.5495584.

[7] G. Dirlich, L. Vogl, M. Plaschke and F. Strian. Cardiac field effects on the EEG. Electroencephalography and Clinical Neurophysiology, 1997. doi: 10.1016/S0013-4694(96)96506-2.

[8] M. Gupta, J. Gao, C. C. Aggarwal and J. Han. Outlier Detection for Temporal Data: A Survey. In: IEEE Transactions on Knowledge and Data Engineering. 2014. doi: 10.1109/TKDE.2013.184.

[9] C. C. Aggarwal. Outlier analysis. 2013.

[10] K. Kourou, T. P. Exarchos, K. P. Exarchos, M. V. Karamouzis and D. I. Fotiadis. Machine learning applications in cancer prognosis and prediction. Computational and Structural Biotechnology Journal, 2015. doi: 10.1016/j.csbj.2014.111.005.

[11] N. Duta and M. Sonka. Segmentation and interpretation of MR brain images: An improved active shape model. In: IEEE Transactions on Medical Imaging, 1998, doi: 10.1109/42.746716.

[12] Y. Zheng, S. Lin, C. Kambhamettu, J. Yu and S. B. Kang. Single-image vignetting correction. In: IEEE Transactions on Pattern Analysis and Machine Intelligence, 2009, doi: 10.1109/TPAMI.2008.263.
[13] S. Zair, S. Le Hégarat-Mascle and E. Seignez. A-contrario modeling for robust localization using raw GNSS data. In: IEEE Transactions on Intelligent Transportation Systems, 2016, doi: 10.1109/TITS.2015.2502279.
[14] D. G. Yoo, S. J. Kang and Y. Hwan Kim. Direction-select motion estimation for motion-compensated frame rate UP-conversion. Journal of Display Technology, 2013. doi: 10.1109/JDT.2013.2263374.
[15] T. Shon and J. Moon. A hybrid machine learning approach to network anomaly detection. Information Science (Ny), 2007. doi: 10.1016/j.ins.2007.03.025.
[16] X. Goldberg. Introduction to semi-supervised learning. Synthesis Lectures on Artificial Intelligence and Machine Learning, 2009. doi: 10.2200/S00196ED1V01Y200906AIM006.
[17] J. C. Munson and T. M. Khoshgoftaar. Regression modelling of software quality: Empirical investigation. Information and Software Technology, 1990. doi: 10.1016/0950-5849(90)90109-5.
[18] Y. Zhang, N. Meratnia and P. Havinga. Outlier detection techniques for wireless sensor networks: A survey. IEEE Commun. Surv. Tutorials, 2010, doi: 10.1109/SURV.2010.021510.00088.
[19] C. C. Aggarwal and C. C. Aggarwal. An Introduction to Outlier Analysis. Outlier Analysis, 2017.
[20] E. C. Erkuş and V. Purutçuoğlu. Outlier detection and quasi-periodicity optimization algorithm: Frequency domain based outlier detection (FOD). European Journal of Operational Research, 2020. doi: 10.1016/j.ejor.2020.011.014.
[21] J. Henriques, et al. Outliers detection in network services with self-learned profiles. In: International Congress on Ultra Modern Telecommunications and Control Systems and Workshops, 2017, doi: 10.1109/ICUMT.2017.8255153.
[22] J. S. Bendat and A. G. Piersol. Random Data: Analysis and Measurement Procedures. Fourth Edition, 2012.
[23] A. Quarteroni, L. Formaggia and A. Veneziani. Cardiovascular Mathematics: Modeling and simulation of the circulatory system. Modeling and Simulation of Applications, 2009.
[24] F. Kusumoto. ECG interpretation: From pathophysiology to clinical application. 2020.
[25] R. J. Oweis and B. O. Al-Tabbaa. QRS detection and heart rate variability analysis: A survey. Biomedical Engineering, 2014.
[26] J. C. Hsieh, W. C. Tzeng, Y. C. Yang and S. M. Shieh. Detecting ECG characteristic points by novel hybrid wavelet transforms: An evaluation of clinical SCP-ECG database. Computers in Cardiology, 2005. doi: 10.1109/CIC.2005.1588213.
[27] M. Aqil, A. Jbari and A. Bourouhou. Evaluation of time-frequency and wavelet analysis of ECG signals. In: Proceedings of 2015 IEEE World Conference on Complex Systems, WCCS 2015, 2016, doi: 10.1109/ICoCS.2015.7483229.
[28] U. R. Acharya, K. P. Joseph, N. Kannathal, C. M. Lim and J. S. Suri. Heart rate variability: A review. Medical and Biological Engineering and Computing, 2006. doi: 10.1007/s11517-006-0119-0.
[29] S. H. Jambukia, V. K. Dabhi and H. B. Prajapati. Classification of ECG signals using machine learning techniques: A survey. In: Conference Proceeding – 2015 International Conference on Advances in Computer Engineering and Applications, ICACEA 2015, 2015, doi: 10.1109/ICACEA.2015.7164783.
[30] V. Seena and J. Yomas. A review on feature extraction and denoising of ECG signal using wavelet transform. In: Proceedings of the IEEE International Caracas Conference on Devices, Circuits and Systems, ICCDCS, 2014, doi: 10.1109/ICDCSyst.2014.6926190.
[31] M. S. Manikandan and K. P. Soman. A novel method for detecting R-peaks in electrocardiogram (ECG) signal. Biomedical Signal Processing and Control, 2012. doi: 10.1016/j.bspc.20111.03.004.
[32] H. P. Kew and D. U. Jeong. Variable threshold method for ECG R-peak detection. Journal of Medical Systems, 2011. doi: 10.1007/s10916-011-9745-7.

[33] L. Sathyapriya, L. Murali and T. Manigandan. Analysis and detection R-peak detection using Modified Pan-Tompkins algorithm. In: Proceedings of 2014 IEEE International Conference on Advanced Communication, Control and Computing Technologies, ICACCCT 2014, 2015, doi: 10.1109/ICACCCT.2014.7019490.

[34] D. Sadhukhan and M. Mitra. R-Peak Detection Algorithm for ECG using Double Difference And RR Interval Processing. Procedia Technology, 2012. doi: 10.1016/j.protcy.2012.05.143.

[35] S. Sahoo, P. Biswal, T. Das and S. Sabut. De-noising of ECG signal and QRS detection using hilbert transform and adaptive thresholding. Procedia Technology, 2016. doi: 10.1016/j.protcy.2016.08.082.

[36] T. Chanwimalueang, W. Von Rosenberg and D. P. Mandic. Enabling R-peak detection in wearable ECG: Combining matched filtering and Hilbert transform. In: International Conference on Digital Signal Processing, DSP, 2015, doi: 10.1109/ICDSP.2015.7251845.

[37] C. Li, C. Zheng and C. Tai. Detection of ECG Characteristic Points Using Wavelet Transforms. In: IEEE Transactions on Biomedical Engineering, 1995, doi: 10.1109/10.362922.

[38] I. Nouira, A. Ben Abdallah, M. H. Bedoui and M. Dogui. A robust R peak detection algorithm using wavelet transform for heart rate variability studies. International Journal on Electrical Engineering and Informatics, 2013. doi: 10.15676/ijeei.2013.5.3.3.

[39] H. Y. Lin, S. Y. Liang, Y. L. Ho, Y. H. Lin and H. P. Ma. Discrete-wavelet-transform-based noise removal and feature extraction for ECG signals. IRBM, 2014. doi: 10.1016/j.irbm.2014.10.004.

[40] J. Laitala et al. Robust ECG R-peak detection using LSTM. In: Proceedings of the ACM Symposium on Applied Computing, 2020, doi: 10.1145/3341105.3373945.

[41] H. Chen and K. Maharatna. An Automatic R and T Peak Detection Method Based on the Combination of Hierarchical Clustering and Discrete Wavelet Transform. In: IEEE Journal of Health & Biomedical Informatics, 2020, doi: 10.1109/JBHI.2020.2973982.

[42] Y. Madhu Keerthana and M. Kiran Reddy. Detecting R-Peaks in electrocardiogram signal using Hilbert envelope. Advances in Intelligent Systems and Computing, 2019. doi: 10.1007/978-981-13-0923-6_1.

[43] V. J. Hodge and J. Austin. A survey of outlier detection methodologies. Artificial Intelligence Review, 2004. doi: 10.1023/B:AIRE.0000045502.109411.a9.

[44] J. Tolvi. Outliers in time series: A review. Journal of the Royal Statistical Society: Series B (Statistical Methodology), 1998.

[45] M. Kantardzic. Data Mining: Concepts, Models, Methods, and Algorithms. Second Edition, 2011.

[46] E. Parzen. 11. Random Variables and Stochastic Processes. Stochastic Processes, 1999.

[47] P. Mishra, C. M. Pandey, U. Singh, A. Gupta, C. Sahu and A. Keshri. Descriptive statistics and normality tests for statistical data. Annals of Cardiac Anaesthesia, 2019. doi: 10.4103/aca.ACA_157_18.

[48] R. Sameni, M. B. Shamsollahi, C. Jutten and G. D. Clifford. A non-linear Bayesian filtering framework for ECG denoising. In: IEEE Transactions on Biomedical Engineering, 2007, doi: 10.1109/TBME.2007.897817.

[49] D. T. Larose and C. D. Larose. Discovering Knowledge in Data: An Introduction to Data Mining. Second Edition, 2014.

[50] P. J. Rousseeuw and M. Hubert. Robust statistics for outlier detection. Wiley Interdisciplinary Reviews: Data Mining and Knowledge Discovery, 2011. doi: 10.1002/widm.2.

[51] G. Sprint, D. J. Cook and M. Schmitter-Edgecombe. Unsupervised detection and analysis of changes in everyday physical activity data. Journal of Biomedical Informatics, 2016. doi: 10.1016/j.jbi.2016.07.020.

[52] E. C. Erkuş, V. Purutçuoğlu and E. Purutçuoğlu. Detection of abnormalities in heart rate using multiple Fourier transforms. International Journal of Environmental Science and Technology, 16(9):2019. doi: 10.1007/s13762-019-02252-3.

[53] O. Yildirim, U. B. Baloglu, R. S. Tan, E. J. Ciaccio and U. R. Acharya. A new approach for arrhythmia classification using deep coded features and LSTM networks. Computer Methods and Programs in Biomedicine, 2019. doi: 10.1016/j.cmpb.2019.05.004.

[54] K. Lan, D. Tong Wang, S. Fong, L. Sheng Liu, K. K. L. Wong and N. Dey. A survey of data mining and deep learning in bioinformatics. Journal of Medical Systems, 2018. doi: 10.1007/s10916-018-1003-9.

[55] F. Andreotti, O. Carr, M. A. F. Pimentel, A. Mahdi and M. De Vos. Comparing feature-based classifiers and convolutional neural networks to detect arrhythmia from short segments of ECG. Computing in Cardiology, 2017. doi: 10.22489/CinC.2017.360-239.

[56] P. Esling and C. Agon. Time-series data mining. ACM Computing Surveys, 2012. doi: 10.1145/2379776.2379788.

[57] S. Matteoli, M. Diani and G. Corsini. A tutorial overview of anomaly detection in hyperspectral images. In: IEEE Aerospace and Electronic Systems Magazine, 2010, doi: 10.1109/MAES.2010.5546306.

[58] P. Berkhin. A survey of clustering data mining techniques. Grouping Multidimensional Data: Recent Advances in Clustering, 2006.

[59] T. C. Fu. A review on time series data mining. Engineering Applications of Artificial Intelligence, 2011. doi: 10.1016/j.engappai.2010.09.007.

[60] K. G. Mehrotra, C. K. Mohan and H. Huang. Anomaly Detection Principles and Algorithms. 2017.

[61] B. Schölkopf, A. Smola and K. R. Müller. Nonlinear Component Analysis as a Kernel Eigenvalue Problem. Neural Computation, 1998. doi: 10.1162/089976698300017467.

[62] A. Duraj. Outlier detection in medical data using linguistic summaries. In: Proceedings – 2017 IEEE International Conference on INnovations in Intelligent SysTems and Applications, INISTA 2017, 2017, doi: 10.1109/INISTA.2017.8001191.

[63] M. Gupta, J. Gao, C. Aggarwal and J. Han. Outlier Detection for Temporal Data. Synthesis Lectures on Data Mining and Knowledge Discovery, 2014. doi: 10.2200/s00573ed1v01y201403dmk008.

[64] R. V. Andreão, B. Dorizzi and J. Boudy. ECG signal analysis through hidden Markov models. In: IEEE Transactions on Biomedical Engineering, 2006, doi: 10.1109/TBME.2006.877103.

[65] I. Vasheghani Farahani, A. Chien, R. E. King, M. G. Kay and B. Klenz. Time series anomaly detection from a Markov chain perspective. In: Proceedings – 18th IEEE International Conference on Machine Learning and Applications, ICMLA 2019, 2019, doi: 10.1109/ICMLA.2019.00170.

[66] O. Salem, A. Guerassimov, A. Mehaoua, A. Marcus and B. Furht. Anomaly detection in medical wireless sensor networks using SVM and linear regression models. International Journal of E-Health and Medical Communications, 2014. doi: 10.4018/ijehmc.2014010102.

[67] R. J. Ellis, B. Zhu, J. Koenig, J. F. Thayer and Y. Wang. A careful look at ECG sampling frequency and R-peak interpolation on short-term measures of heart rate variability. Physiological Measurement, 2015. doi: 10.1088/0967-3334/36/9/1827.

[68] V. Chandola, A. Banerjee and V. Kumar. Outlier detection: A survey. ACM Computing Surveys, 2007.

[69] Outlier Detection: Applications And Techniques. International Journal of Computer Science Issues, 2012.

[70] S. Basu and M. Meckesheimer. Automatic outlier detection for time series: An application to sensor data. Knowledge and Information Systems, 2007. doi: 10.1007/s10115-006-0026-6.

[71] F. Strasser, M. Muma and A. M. Zoubir. Motion artifact removal in ECG signals using multi-resolution thresholding. In: European Signal Processing Conference, 2012.

[72] Z. Zhao, Y. Zhang, Z. Comert and Y. Deng. Computer-aided diagnosis system of fetal hypoxia incorporating recurrence plot with convolutional neural network. Frontiers in Physiology, 2019. doi: 10.3389/fphys.2019.00255.

[73] J. S. Richman and J. R. Moorman. Physiological time-series analysis using approximate and sample entropy. American Journal of Physiology-Heart and Circulatory Physiology, 2000. doi: 10.1152/ajpheart.2000.278.6.h2039.

[74] K. H. Ong, D. Ramachandram, R. Mandava and I. L. Shuaib. Automatic white matter lesion segmentation using an adaptive outlier detection method. Magnetic Resonance Imaging, 2012. doi: 10.1016/j.mri.2012.011.007.

[75] M. Prastawa, E. Bullitt, S. Ho and G. Gerig. A brain tumor segmentation framework based on outlier detection. Medical Image Analysis, 2004. doi: 10.1016/j.media.2004.06.007.

[76] J. Lee, D. D. McManus, S. Merchant and K. H. Chon. Automatic motion and noise artifact detection in Holter ECG data using empirical mode decomposition and statistical approaches. In: IEEE Transactions on Biomedical Engineering, 2012, doi: 10.1109/TBME.20111.2175729.
[77] A. R. Magnano, S. Holleran, R. Ramakrishnan, J. A. Reiffel and D. M. Bloomfield. Autonomic modulation of the U wave during sympathomimetic stimulation and vagal inhibition in normal individuals. PACE – Pacing and Clinical Electrophysiology, 2004. doi: 10.1111/j.1540-8159.2004.00665.x.
[78] N. Heim and J. E. Avery. Adaptive anomaly detection in chaotic time series with a spatially aware echo state network. arXiv, 2019.
[79] G. E. P. Box and D. A. Pierce. Distribution of residual autocorrelations in autoregressive-integrated moving average time series models. Journal of the American Statistical Association, 1970. doi: 10.1080/01621459.1970.10481180.
[80] A. Sagoolmuang and K. Sinapiromsaran. Median-difference window subseries score for contextual anomaly on time series. In: 2017 8th International Conference on Information and Communication Technology for Embedded Systems, IC-ICTES 2017 – Proceedings, 2017, doi: 10.1109/ICTEmSys.2017.7958772.
[81] M. Munir, S. Erkel, A. Dengel and S. Ahmed. Pattern-based contextual anomaly detection in HVAC systems. In: IEEE International Conference on Data Mining Workshops, ICDMW, 2017, doi: 10.1109/ICDMW.2017.150.
[82] J. Wang, R. Li, R. Li and B. Fu. A knowledge-based deep learning method for ECG signal delineation. Future Generation Computer Systems, 2020. doi: 10.1016/j.future.2020.02.068.
[83] W. Midani, Z. Fki and M. Benayed. Online anomaly detection in ECG signal using Hierarchical Temporal Memory. In: International Conference on Advances in Biomedical Engineering, ICABME, 2019, doi: 10.1109/ICABME47164.2019.8940307.
[84] C. Dora and P. K. Biswal. Robust ECG artifact removal from EEG using continuous wavelet transformation and linear regression. In: 2016 International Conference on Signal Processing and Communications, SPCOM 2016, 2016, doi: 10.1109/SPCOM.2016.7746620.
[85] E. C. Erkuş and V. Purutçuoğlu. Outlier detection and quasi-periodicity optimization algorithm: Frequency domain based outlier detection (FOD). European Journal of Operational Research, 2020. doi: 10.1016/j.ejor.2020.011.014.
[86] J. Ye, T. Kobayashi, M. Murakawa, T. Higuchi and N. Otsu. Anomaly detection using multi-channel FLAC for supporting diagnosis of ECG. IEEJ Transactions on Electronics, Information and Systems, 2012. doi: 10.1541/ieejeiss.132.111.
[87] J. R. Pinto, J. S. Cardoso, A. Lourenço and C. Carreiras. Towards a continuous biometric system based on ECG signals acquired on the steering wheel. Sensors (Switzerland), 2017. doi: 10.3390/s17102228.
[88] Z. Chen and Y. F. Li. Anomaly detection based on enhanced DBScan algorithm. Procedia Engineering, 2011. doi: 10.1016/j.proeng.20111.08.036.
[89] R. Varatharajan, G. Manogaran and M. K. Priyan. A big data classification approach using LDA with an enhanced SVM method for ECG signals in cloud computing. Multimedia Tools and Applications, 2018. doi: 10.1007/s11042-017-5318-1.
[90] J. G. Webster, T. Andrade, J. Gama, R. P. Ribeiro, W. Sousa and A. Carvalho. Anomaly detection in sequential data: Principles and case studies. Wiley Encyclopedia of Electrical and Electronics Engineering, 2019.
[91] S. Elmougy, M. S. Hossain, A. S. Tolba, M. F. Alhamid and G. Muhammad. A parameter based growing ensemble of self-organizing maps for outlier detection in healthcare. Cluster Computing, 2019. doi: 10.1007/s10586-017-1327-0.
[92] N. Peters, M. A. Gatzoulis and R. Vecht. ECG Diagnosis in Clinical Practice. 2009.
[93] R. Chalapathy and S. Chawla. Deep learning for anomaly detection: A survey. arXiv, 2019.
[94] F. M. Nolle, F. K. Badura, J. M. Catlett, R. W. Bowser and M. H. Sketch. CREI-GARD, A new concept in computerized arrhythmia monitoring systems. Computers in Cardiology, 1987.

[95] F. Andreotti, J. Behar, S. Zaunseder, J. Oster and G. D. Clifford. An open-source framework for stress-testing non-invasive foetal ECG extraction algorithms. Physiological Measurement, 2016. doi: 10.1088/0967-3334/37/5/627.

[96] V. Behravan, N. E. Glover, R. Farry, P. Y. Chiang and M. Shoaib. Rate-adaptive compressed-sensing and sparsity variance of biomedical signals. In: 2015 IEEE 12th International Conference on Wearable and Implantable Body Sensor Networks, BSN 2015, 2015, doi: 10.1109/BSN.2015.7299419.

Jabulani Monchwe, Elias Munapo and Martin Chanza

Chapter 12
Optimization of debt collection strategies for South African banks with machine learning models

Abstract: Once banks are no longer able to collect money from defaulted accounts, they turn to third-party debt collectors to recover the funds. These debt collectors use aggressive tactics, such as contacting the debtor's family and friends, to try to locate the money. Because of this, many people have become afraid of ever becoming delinquent on their loan payments again. This has led to a decrease in the number of people who are able to become debt-free.

This chapter presents machine learning models for optimization debt collection for South African banks. The main objective of this study is to compare the accuracy and cost optimization of logistic regression (LR), random forest (RF), multilayer perceptron (MLP) and decision tree (DT) models using bank delinquency data. Findings reveal that the LR and RF classifiers achieved the highest debt recovery rates. Since the LR model only considers SMS as a communication mode compared to RF, which has diverse communication modes, the RF model is selected as the optimal model to prescribe communication strategies. Communication strategies are based on classifiers that have the highest Hamming score, high subset accuracy, lowest action cost and diversity of communication modes.

Keywords: Machine learning, multilabel classification, communication treatment strategy, debt collection recoveries

12.1 Introduction

According to Romeo and Sandler [1], debt collection is one of the fastest-growing areas in South Africa but has only received minimal attention from quantitative analysts and machine learning (ML) practitioners. Mothibi [2] also states that debt levels in South Africa were reported to have escalated to 48.8% of the gross domestic product in 2017, causing major concern and strain on the collection industry and government leadership As the value of unpaid debts increases, intervention remains minimal. Romeo and

Jabulani Monchwe, Martin Chanza, Department of Statistics and Operations Research, School of Economic Sciences, North-West University, Mafikeng, South Africa
Elias Munapo, Department of Statistics and Operations Research, School of Economic Sciences, North-West University, Mafikeng, South Africa, e-mail: Elias.Munapo@nwu.ac.za

https://doi.org/10.1515/9783110717655-012

Sandler [1] reported that financial institutions, including banks, face tough decisions to write off millions of unpaid debts every year, knowing that not all debts will be paid back on time. Faced with this problem, banks are forced to hand over their debt to external collection agencies or forfeit possible future repayments. According to Van de Geer et al. [3], Mitcher and Peterson were the first to look at the problem of optimizing debt collection for various portfolios in 1957 Their focus was to determine the duration for which the debt collection agent should pursue the debtor for repayment. This optimization method resulted in a 33% improvement in recovery rates, highlighting the optimization of optimization methods for debt collection agencies.

Dias et al. [4] acknowledged that recovering delinquent debt payments is incredibly challenging for collection agents, as they must determine the relevant communication mode or treatment strategy suited for the debtor in question. For example, being polite to a good payer, aggressive to consistent defaulters, sympathizing to good-paying debtors currently facing short-term financial problems, and knowing when to execute legal actions are very important. Determining the relevant treatment strategy to apply for each collection attempt is not ideal for agents, as they do not have the necessary skills or resources to compute the optimal communication treatment strategies. This can cause the debt collection company to lose important debtors due to inappropriate treatments and can also result in unnecessary operational costs [4]. So et al. [5] suggested that for a company to minimize costs and boost collection rates, an effective treatment strategy should include all actions, namely: gentle reminder actions (i.e., Emails, short message service (SMS) and letters), tough actions (fixed letters and aggressive phone calls) and stricter actions (legal channels or property recovery). So et al. [5] further state that an optimal treatment channel should decide which of the treatments to action, for how long and in what sequence, given the state of the account.

Since Phillips and Moggridge [6] defined ML as a discipline that combines knowledge of mathematics, statistics, programming and Optimization, it also means that ML can be used to solve both supervised (regression and classification tasks) and unsupervised (dimension reduction and clustering) learning problems. Carleo et al. [7] support that supervised ML can be used in learning optimized classes and applying them to new datasets, which could result in more than just predicted classes but rather optimized predicted classes. Hence, the supervised ML models will be used in the study to learn optimized classes, and the models include decision tree (DT), random forest (RF), logistic regression (LR) and multilayer perceptron (MLP).

Wang et al. [8] define a DT model as a technique that can be used for both classification and regression tasks. The idea of the model is to generate decision rules based on attributes found in the data sample. Aroef et al. [9] describe the DT algorithm using three levels of nodes, namely the parent node (root), child node (internal) and leaf node (child without branch). Aroef et al. [9] also state that Ho introduced RF in 1995, aiming to overcome the problem of splitting nodes and ensemble learning of DT using bootstrap to resample the dataset and optimally select features. In the study by

Breiman in 2001, as cited in [10], RF models are tree-based algorithms that follow a rule-based approach to training individual trees and removing high correlations.

According to Potdar et al. [11], the MLP is a supervised deep learning (DL) algorithm that forms a solid foundation for most DL applications across industries. The MLP model can be useful for both regression and classification tasks. The architecture of this learner uses interconnected nodes to process and understand the information provided. The advantage of this learner lies in its ability to self-update and adapt its network structure to model complex relationships that map the input features to the desired outputs, as given in [11]. The study by Fan et al. in 2008, as cited in [12], states that LR models are probabilistic because they are intended to return the maximum likelihood estimation and use a logit-link function to analyze and establish the relationships between the explanatory and the explained variables.

Considering the above background and the evidence presented, this research will seek to compare the LR, RF, MLP and DT models in optimizing debt collection strategies for banking portfolios. This study will develop a Treatment Channel model that ensures debtors are treated with appropriate actions and that action costs are optimized. This model will be developed to prescribe the relevant communication strategies.

12.2 Related literature

This part of the study examines the preliminary research performed in the context of debt collection, ML modeling and treatment channel strategies. The section briefly explores the problems faced by previous research, the scientific methodologies and models applied, data-driven results, limitations uncovered and recommended ways to address them. The purpose of this section is to pave the way for future studies related to collection strategy optimization using statistical modeling techniques and to provide a strategy to overcome the limitations faced by previous research practitioners over time. The section will review studies where treatment channel models are applied to prescribe the next best actions in the context of debt collection.

Shoghi [13] conducted a study that applied prescriptive analytics and optimization methods to improve the debtor contact scheduling policy. This policy feeds into the treatment communication channels for each debtor in a debt collection firm. The main goal was to reduce costs and increase productivity by determining which debtors to call and on which day, given the maximum call constraints for the day [13]. The data was obtained by conducting a controlled experiment for an insurance company in the Netherlands. The variables used were split into two categories: Debtor's Historical Interactions and Debtor behavior. For debtor behavior, the following factors were considered: amount owed, number of partial payments in the past, existing promise to pay (PTP), availability of a phone number, availability of a mobile number, industry, amount already paid, previous payment history (binary outcome), average income inferred from postal code and compared to age group, and receipt of section 29 (letter of final warning). For historical interactions, the following variables were considered: answered phone calls, existing PTP, number of efforts per collector, number of previous calls, number of days since PTP, number of days since the last collector's effort, number of days since right party contact, number of days since the last incoming contact and number of days since the last incoming SMS and call [13].

The algorithms used were the Markov decision process (MDP) to register the probability states, and LightGBM was then used to estimate the value function approximation for the next best action. The results showed an improvement of 14% in the debt collected and a reduction of 22% in calling costs. The shortfall of this study is the inability to handle a larger portfolio of debtors that are nonhomogeneous in nature, which results in high dimensionality and the removal of the email contact method, as shown by Shoghi [13] in Figure 12.1.

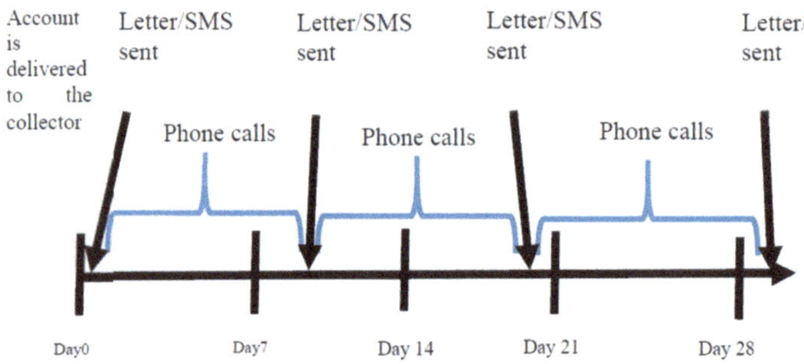

Figure 12.1: Communication treatment channel – structure (source: [14]).

The study will address the curse of dimensionality and include the email contact method in the results.

Drozd and Serrano-Padial [15] conducted a study focusing on the consumer debt industry and debt collection channels. The aim of the study was to understand the factors that lead to higher collection channel costs. The training sample used was collected from the credit bureau of delinquent customers for the period 2000–2006, and the dataset included economic variables that explain the consumer standard of living. Statistical methods were used to understand the cost-benefit analysis, and Markov chain processes were also deployed on the sample base [15]. The evidence presented showed that economic conditions can lead to delinquency, resulting in higher collection costs that can threaten a firm with bankruptcy.

In [16], the study focused on evaluating stochastic repayment behaviors of delinquent debtors for credit card accounts. The aim of the study was to maximize the total repayment amount and minimize the collection costs. The training sample was collected from the Standard Bank Group and contained 6,600 accounts with a total of $6 million in debt, which were handed over to an external debt collector for the period 2004–2006 [16]. The variables considered were owed balance, FICO score, days on book, number of payments, phase 1 commission, transition to the second collection phase and phase 2 commission rate. The methods used were the Generalized Method of Moments and Maximum Likelihood Estimation, with additions of Markov chain processes. Significant improvements were evident between the model outputs and the standard procedure [16]. The treatment score was evaluated in combination with the repayment score, which made it difficult to measure the true performance. The study will separate the two scores and evaluate them independently.

A dynamic stochastic repayment behavior model for delinquent debtors was developed [17]. The aim was to control the expenditure of account treatment actions and only apply them to accounts below a predetermined threshold. The models employed were stochastic optimization and controlled Hawkes process (both of which run a Markov Process in the backend). With the optimized treatment channels, collection costs were reduced from 46.9% to 8.7% of all costs [17]. The shortfall of the study is that the treatment actions only work up until the fourth payment is received; going forward, the account gets removed from the treatment plan. To address this limitation, the research will implement a treatment model that will not only be limited to the number of payments made but will work until the repayment amount is fully recovered.

The study [18] was conducted to optimize debt collection recoveries using Survival Regression. The aim was to minimize operating expenses on debtors who are already paying. To achieve this, the research developed a treatment channel model called the soft collection strategy. The analogy used in the study compares the treatment channel model to medication issued and repayment to the patient's recovery status [18]. The training set was obtained from a Czech bank and collected for the period 2017–2019. The raw dataset contains 43,382 debtor details with up to 30 features. The features were categorized into three groups: product details, debtors' details and recovery process details. The product details bucket includes the following: product class, amount owed, branch (loan obtained), created date, product data and frame, credit card categories, product currency, status and interest rate. The debtors' personal details include the following fields: account open date, employment information, legal state of debtor and creditor, residential details, risk categories, payment history and type of credit with portfolio segment. The recovery process includes the following fields: telephone score, reminder date, delinquent count (in days), call date, portfolio entry details, milestone information and payment history since the transition [18]. The models compared are LR and several survival models in estimating the effect of calling a debtor to maximize the probability of repayment. However, the study was inconclusive as the performance of different models depended on the training data size.

In the existing literature on debt collection recoveries, the research output is minimal, especially regarding treatment channel model applications. The literature also shows that most studies compare ML models to statistical models, and the results mainly depend on the data preparation processes rather than the model techniques themselves. This study aims to improve model accuracy using statistical learning for better data preparation and then compare multiple models. The reviewed models' performances have above-average results; however, with the involvement of ML and recent developments, ML systems can self-direct and discard irrelevant features during the learning process [19]. ML models can also analyze multiple objectives instead of binary strategic option outcomes [20]. This study is important because it uses multiple objective scenarios, making the results more effective. This is the standout point of the study.

12.3 Research methodology

This section describes the ML algorithms, model structures, data tables, Python functions and pseudocode used for the data analysis. Python 3.9 with Jupyter Notebook is used together with other dependency packages such as Pandas, NumPy, Seaborn, Scikit-learn, Plotly and TPOT for genetic optimization. Figure 12.2 displays the model methodology diagram.

12.3.1 Study population and sampling strategy

The overall secondary dataset was sourced from the reporting team of Prime Investigations & Recovery (Ltd) Pty. for the period March–July 2020. The dataset contains 3,124,365 observations with 60 variables. The dataset will be split into three sets: the Validation dataset, which will only contain the July data; the Training dataset, which is 70% of the remainder (from March to June 2020); and the Test dataset, which is 30% of the remainder. To split the remainder set into training and test sets, stratified random sampling will be used to ensure balanced splits between the training and test datasets. Bhardwaj (2019) defines stratified random sampling as a technique that first divides the population into subgroups known as strata. The strata are based on the similarity of characteristics per each observation, and then from each stratum, the members or observations are selected randomly. Further exclusion criteria will be applied to observations that do not add value to the business, such as the removal of handover balances less than R100.00 (<100), account ages less than 0 (<0), and customer ages less than 0 (<0). The business has attached cleaning rules to be applied, such as minimum handover balance, age of the account and customer age, and these rules will be used to filter for the clean dataset.

To ensure that the dataset is statistically clean, normalized and feature-scaled, the following statistical methodologies will be applied: Mean imputation to fill missing values, interquartile range (IQR) to detect outliers, Shapiro-Wilk statistic to test for normality and MinMax scaler to standardize the data records. Once the dataset is cleaned, it will be divided into three sets, namely validation, training and test datasets. The training and test datasets will be subset into independent variables (X/decision drivers) and dependent variables (Y/communication treatment strategies), and

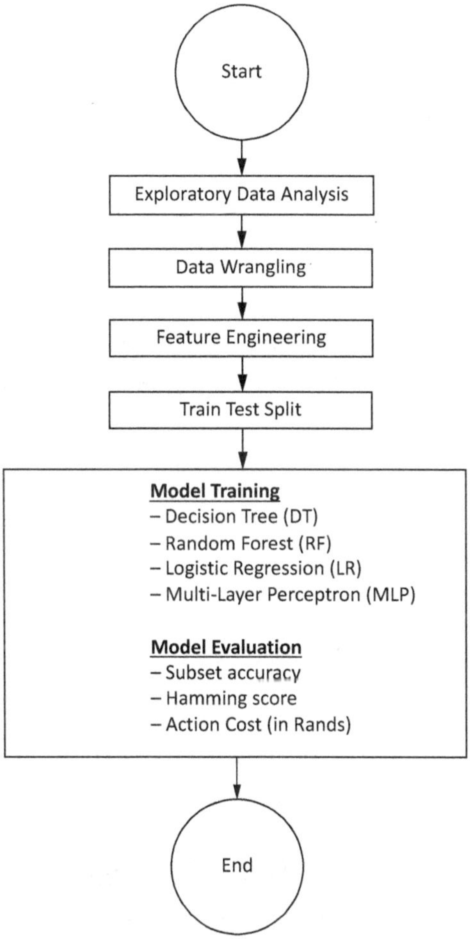

Figure 12.2: Methodological flow.

the supervised multilabel classifier models will be fitted using the training dataset. The ML models that will be fitted are DT, RF, MLP and LR, and the test and validation datasets will be used to evaluate the models' predictive power using the Hamming score, subset accuracy and overall action cost in rands. The model with the highest Hamming score, highest subset accuracy and lowest action cost (with the diversity of treatment actions) will be selected as the best-performing prescriptive model.

12.3.1.1 Mean imputation

According to Qian and Chu [21], as shown in eq. (12.1), mean imputation is a better and more efficient strategy for replacing missing values:

$$\text{Mean} = \frac{\sum_{n=1}^{k} x_n}{k} \tag{12.1}$$

where x is the array of observations, n is the entry in an array and k is the number of observations.

12.3.1.2 Interquartile range (IQR)

According to Mishra et al. [22], the IQR is a statistical method that is very useful in detecting outliers in a variable. This study will use this technique to detect the presence of outliers in the dataset, as follows:

$$\text{IQR} = \frac{Q3 - Q1}{2} \tag{12.2}$$

$$Q1 = \frac{1}{4}(n+1)\text{th term} \tag{12.3}$$

$$Q3 = \frac{3}{4}(n+1)\text{th term} \tag{12.4}$$

where n is the sample size.

12.3.1.3 Shapiro-Wilk test

The null hypothesis will be defined as: The sample comes from a normal distribution. The formal test statistic to be employed will be the Shapiro-Wilk test, to determine if the sample is indeed from a normally distributed population, as follows:

$$W = \frac{\left(\sum_{i=1}^{n} a_i x_{(i)}\right)^2}{\sum_{i=1}^{n} (x_i - \overline{x})^2} \tag{12.5}$$

where x_i is the ith smallest number in the sample\msemicolon for $\overline{x}$, see eq. (12.4), the coefficients $a_{i'} = (a_1, \ldots, a_n) = \frac{m^T V^{-1}}{\left(m^T V^{-1} V^{-1} m\right)^{1/2}}$ $m = (m_1, \ldots, m_n)^T$, vector that made expected values of order statistics of independently and identically distributed random samples from standard normal distribution V is the covariance matrix of those order statistics.

The Shapiro-Wilk test will return a p-value that will be compared against an alpha value of 5% (as a rule of thumb). If the p-value is less than 5%, then the null hypothesis will be rejected. The studies done by [23–29] reveal alternative scalar methods in eq. (12.5):

Parametric transformers

1. Standard scaler $= \dfrac{x_i - \mu}{\sigma}$

2. Normalizer $= \dfrac{x_i - x_{\min}}{x_{\max} - x_{\min}}$

3. Robust scaler $= \dfrac{x_i - Q2}{Q3 - Q1}$

4. Max Abs scaler $= \dfrac{x_i}{\max(|x|)}$

5. MinMax scaler $= \dfrac{x_i - x_{min}}{x_{max} - x_{min}}.(\max - \min) + \min;\ \text{defaults}(\min = 0, \max = 1)$

Nonparametric transformers

6. Power transformer(Yeo-Johnson) $= \psi(x_i, \lambda) = \begin{cases} \dfrac{(x_i+1)^{\lambda-1}}{\lambda} & x_i \geq 0 \text{ and } \lambda \neq 0, \\ \log(x_i+1) & x_i \geq 0 \text{ and } \lambda = 0, \\ -\dfrac{(-x_i+1)^{2-\lambda}-1}{2-\lambda} & x_i < 0 \text{ and } \lambda \neq 2, \\ -\log(-x_i+1) & x_i < 0 \text{ and } \lambda = 2, \end{cases}$

7. Quantile transformer $= F_o^{-1}(F_m.P_m)$

8. Polynomial features $= f(y) = b_o + b_1y_1 + b2y_2^2 + \cdots + b_ny_1^n$

where x_i is current observation

$\mu = \text{eq. } (12.4)$

$n = \text{sample size}$

$\sigma = \sqrt{\dfrac{\sum |x_i - \mu|^2}{n}}$

$Q1 = \text{eq. } (12.6)$

$Q2 = \dfrac{2}{4}(n+1)\text{th term}$

$Q3 = \text{eq. } (12.7)$

λ is the estimated using maximum likelihood estimation

$F_o^{-1} = \text{cdf of } F(x) = p(X \leq x) = \int_{-\infty}^{x} (f(t)dt) \text{for } x_i \in R$

$F_m = \text{cdf of } P_m$ and P_m = modeled variable, P_o = observed variable

b_o = regression parameter estimate. (12.6)

The aforementioned methods will be applied until the normality assumption is satisfied.

12.3.1.4 Train_Test-split

The scikit-learn algorithm called Train_Test-split will be used as it applies stratified random sampling in the backend.

12.3.2 Decision tree learner

According to Wang et al. [8], the DT model is a supervised learning technique that can be used for both classification and regression tasks. The idea of the model is to generate decision rules based on attributes found in the training sample set. This kind of ML model generates nodes that each represents characteristics found in the dataset or the environment exposed to it. Aroef et al. [9] describe this algorithm using three levels of nodes, namely the parent node (root), child node (internal) and leaf node (child without branch), as shown in Figure 12.3. The DT learner will be exposed to the sample dataset to extract and create business rules. In the case of improved learning, the genetic algorithm will be used to tune the hyper-parameters. The parameters to be tuned for the DT learner include selection criterion (Gini or entropy), max_depth ranging between 1 and 11, min_samples_split ranging between 2 and 21, and min_samples_leaf ranging between 1 and 21.

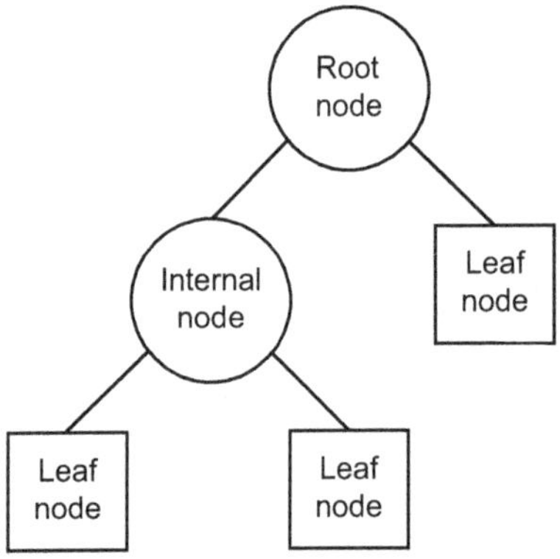

Figure 12.3: DT model structure (source: [9]).

12.3.3 Random forest learner

According to Aroef et al. [9], Ho introduced RF in 1995 with the aim of overcoming the problem of splitting nodes and enhancing the ensemble learning of DT by using bootstrap resampling of the dataset and optimal feature selection. Breiman, in his 2001 study as cited by [10], described the RF model as a tree-based algorithm that follows a rule-based approach to training individual trees and reducing high correlations. This approach produces its predictors in an ensemble manner, where all the trees are voted upon for nodes with a similar structure. To further reduce the correlation between trees and the variance of ensemble predictors, RF use a subset of $m < p$ to randomly vote for their predictors every time there is a split [10].

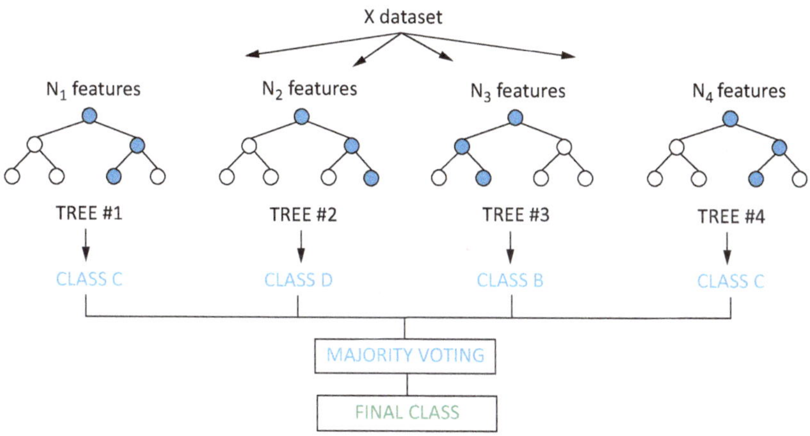

Figure 12.4: RF model structure (source: [30]).

To optimize the performance of this learner, the study shown in Figure 12.4 will employ a genetic algorithm to tune the following parameters: n_estimators = 100, criterion (Gini or entropy), max_features ranging from 0.05 to 1.0 in increments of 0.05, min_samples_split ranging from 2 to 21, min_samples_leaf ranging from 1 to 21, and a Boolean variable for bootstrap, which can be either True or False.

12.3.4 Multilayer perceptron

According to Potdar et al. [11], the MLP is a supervised DL algorithm that forms a solid foundation for most ML applications across industries. The MLP learning model can be useful for both regression and classification tasks. The architecture of this learner uses interconnected nodes to process and understand the information provided. The advantage of this learner lies in its ability to self-update and adapt its network structure to model complex relationships that map the input features to desired outputs. Potdar et al. [11] mention that the factors contributing to the outcome of a learner are influenced or determined by feature characteristics and the weights associated with the network connections derived from input data. According to Seragadam et al. [14], this learner self-updates its weights and accumulates learning experience until it reaches a satisfactory level, as shown in Figure (21).

To assist the learner in optimizing the learning experience, the following figure 12.5 explains hyper-parameters will be tuned: hidden_layer_sizes will default to 100, the activation function will range between identity, logistic, tanh and relu, the solver will range between lbfgs, sgd and adam, the learning_rate will range between constant, invscaling and adaptive, max_iter will be kept constant, the early_stopping option will be set to true to prevent the model from overfitting and the validation_fraction will be explored between 0.001 and 0.1.

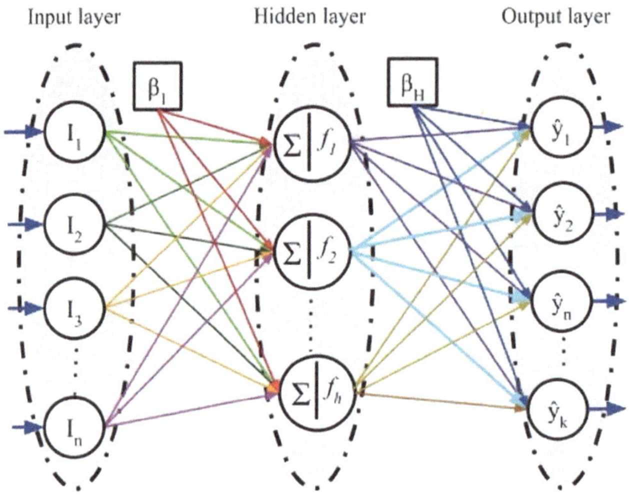

Figure 12.5: MLP model structure (source: [31]).

12.3.5 Logistic regression classifier

According to the study by Fan et al. in 2008, as cited in [12], LR models are probabilistic in nature (intended to return the maximum likelihood estimation), unlike the previously mentioned tree-based and ANN models that are deterministic (intended to predict a class instead of a probability score). The LR learner in Figure 12.6 uses a logit-link function to analyze and establish the relationships between the explanatory and the explained variables. The advantage of this learner lies in its learning method, which is to minimize the negative log-likelihood function in the assignment of the model learning weights and introduce a penalty to the log-likelihood function when maximizing the likelihood estimation, which helps in overcoming the high correlation between features. Like any other, the analysis will hyper-parameter tune the following: the best penalty strategy will be explored using (l1 and l2), CV between (3 and 5), C between (1e-4, 1e-3, 1e-2, 1e-1, 0.5, 1., 5., 10., 15., 20., 25.), and the dual factor will be investigated based on its Boolean structure (True and False).

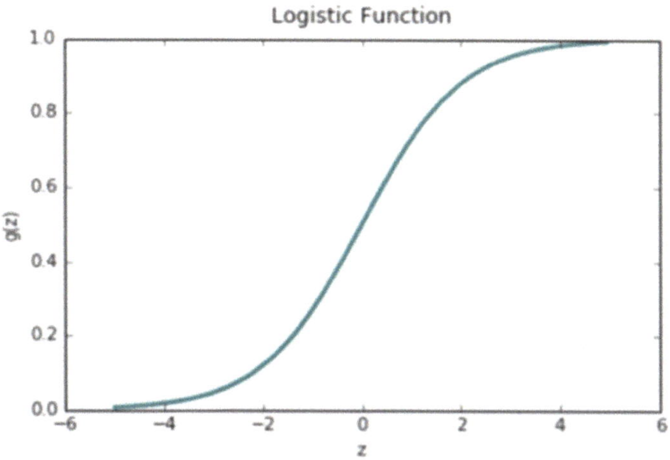

Figure 12.6: LR model structure (source: [30]).

12.3.6 Model selection criterion

3.6.1 Hamming score and subset accuracy

The article [32] defines the Hamming score as a measure that returns the fraction of correctly predicted values for all labels. Reference [33] defines subset accuracy or exact match ratio as a model performance measure that increases when the model correctly classifies every possible label without increasing the false-positive rate. Hamming score and subset accuracy will be used to evaluate the multilabel model outputs and will follow the algorithm shown in Figure 12.7 and the subset of the algorithm written in Figure 12.8.

3.6.2 Cost optimization

Cost analysis is a subset of cost-benefit analysis, which is defined by Harcourt [34] as a tool used by businesses to calculate potential earnings if a specific project is undertaken. Cost analysis will be used in this study as a tool to guide in selecting the cost-optimal model. The formulation includes multiplying the action count by its related cost, as shown in the algorithm below.

```
def MultiClass_MultiLabel_Scoring(y_pred, y_true):
    '''
    The algorithm accepts two dataframes (true label and predicted values)
    And compute the subset and overall scoring metrics for: Hamming & Balanced accuracy
    '''
    true_cols = y_true.columns
    True_Col=[]
    HammingScore=[]
    Accuracy_Score=[]
    Result_DF = pd.DataFrame()

    for i in range(0, y_true.shape[1]):
        pred_col = y_pred[i]
        true_col = y_true[true_cols[i]]
        True_Col.append(true_cols[i])

        from sklearn.metrics import hamming_loss #compute Hamming score
        HammingLoss = (hamming_loss(true_col, pred_col))
        HammingScore.append(1-HammingLoss)

        from sklearn.metrics import balanced_accuracy_score #compute balanced accuracy score
        subset_acc = balanced_accuracy_score(true_col, pred_col)
        Accuracy_Score.append(subset_acc)

    Result_DF['True Columns']=True_Col
    Result_DF['Subset_HammingScore']=HammingScore
    Result_DF['Subset_Bal_Accuracy']=Accuracy_Score

    print('[Overall Haming_Score=' + str(round(Result_DF['Subset_HammingScore'].sum()/len(Result_DF), 4)) + ']'
          , '[Overall Balanced Accuracy=' + str(round(Result_DF['Subset_Bal_Accuracy'].sum()/len(Result_DF), 4)) + ']'
          )
    return Result_DF
```

Figure 12.7: Hamming and balanced accuracy score – algorithm.

12.3.7 Ethical considerations

This study does not include human or animal participation; it focuses solely on the analysis of secondary data sourced from Prime Investigations & Recoveries (Pty) Ltd. using statistical techniques. A permission letter was obtained from the Managing Director of Prime Investigations & Recoveries (Pty) Ltd. The permission letter was obtained and adhered to accordingly. Any information obtained in connection with this study will remain confidential and will not be disclosed in a manner that identifies the source of the data. The data will be protected by ensuring that field names and results are encoded to prevent exposure of any department. The performance metrics highlighted in the data will be kept confidential, and only the researcher and project supervisors will have access to these raw and processed data sheets.

This section provides a detailed discussion of all the datasets, mathematical equations and other measures that will be applied in the analysis of South African banking portfolios. The study takes into consideration all related steps, such as missing values imputation, outlier detection, feature scaling, model fitting and model evaluation. The following section of the study will focus on applying the above-mentioned techniques to actual datasets.

```
def Cost_Analysis(input_df, Call_R, SMS_R, Email_R, start_Pos, End_Pos):
    '''
    This algorithm accepts six arguments,  dataframes (true label and predicted values)
    And computes the subset and overall scoring metrics for: Hamming & Balanced accuracy & Cost in Rands
    '''
    #1. Substring the Communicatiom type
    str_=[]
    for i in range(0, len(input_df)):
        char = input_df['True Columns'].str.slice(start_Pos, End_Pos)[i]
        str_.append(char)
    input_df['string']=str_ .

    #2. Calculate CALL costs for true and predicted labels
    C = input_df[input_df['string'] == 'C']
    C['true_C'] = C['True_actions']*Call_R
    C['pred_C'] = C['Pred_actions']*Call_R

    #3. Calculate SMS costs for true and predicted labels
    S = input_df[input_df['string'] == 'S']
    S['true_C'] = S['True_actions']*SMS_R
    S['pred_C'] = S['Pred_actions']*SMS_R

    #4. Calculate EMAIL costs for true and predicted labels
    E = input_df[input_df['string'] == 'E']
    E['true_C'] = E['True_actions']*Email_R
    E['pred_C'] = E['Pred_actions']*Email_R

    #5. Combine the costs for true and predicted labels
    Costs = pd.concat([C, S, E],axis=0)
    Costs['Savings_R'] = Costs['true_C'] - Costs['pred_C']
    Costs = Costs.reset_index(drop=True)

    print('[Total True_label Action_Costs=R'+str(round(Costs['true_C'].sum(),2)) + ']'
          ,'[Total Pred_label Action_Costs=R'+str(round(Costs['pred_C'].sum(),2)) + ']'
          ,'==>Surplus/Deficit=R' + str(round(Costs['true_C'].sum() - Costs['pred_C'].sum(),2))
    )

    return Costs
```

```
#Call the function & calculate True vs. Predicted Costs
Cost_Analysis(input_df = MultiClass_MultiLabel_Scoring(y_pred=pd.DataFrame(DT_.predict(Input_Features))
                                                      , y_true=y_Base
                                                      )
              , Call_R = 2.5
              , SMS_R = 0.5
              , Email_R = 0.005
              ,start_Pos=7, End_Pos=8
             )
```

Figure 12.8: Subset-action cost – algorithm.

12.4 Data analysis, results and discussion

This section discusses the parameter estimation of selected candidate supervised multilabel models for the communication treatment strategy. The candidate models to be fitted are DT, RF, MLP and LR. The initial models will be fitted using basic model parameters to narrow the selection process. If needed, the hyper-parameter tuning process will follow for the selected models using a genetic algorithm for parameter searching.

12.4.1 Decision tree validation: prescriptive results

The first model to be fitted is a multilabel DT classifier model with estimated parameters: ccp_alpha, min_impurity_decrease, min_weight_fraction_leaf and random_state all equal to 0.0; and class_weight, max_depth, max_features and min_impurity_split all equal to None, criterion = gini, min_samples_leaf = 1, min_samples_split = 2, splitter = best, base estimator = DecisionTreeClassifier and n_jobs = −1, as shown in Figure 12.9.

```
{'estimator__ccp_alpha': 0.0,
 'estimator__class_weight': None,
 'estimator__criterion': 'gini',
 'estimator__max_depth': None,
 'estimator__max_features': None,
 'estimator__max_leaf_nodes': None,
 'estimator__min_impurity_decrease': 0.0,
 'estimator__min_impurity_split': None,
 'estimator__min_samples_leaf': 1,
 'estimator__min_samples_split': 2,
 'estimator__min_weight_fraction_leaf': 0.0,
 'estimator__random_state': 0,
 'estimator__splitter': 'best',
 'estimator': DecisionTreeClassifier(random_state=0),
 'n_jobs': -1}
```

Figure 12.9: DT model parameters.

The model evaluation was performed using a validation dataset, and the results are summarized in Table 12.1.

Table 12.1: Decision tree evaluation results.

```
Candidate Model: MultiOutputClassifier(estimator=DecisionTreeClassifier(random_state=0),
                      n_jobs=-1)
[Overall Haming_Score=0.9649] [Overall Balanced Accuracy=0.1987]
[Total True_label Action_Costs=R438215.75] [Total Pred_label Action_Costs=R546365.38] ==>Surplus/Deficit=R-108149.62
```

	True Columns	Subset_HammingScore	Subset_Bal_Accuracy	True_actions	Pred_actions	string	true_C	pred_C	Savings_R
0	PAY_W1_C	0.988305	0.344038	3187.0	7988.0	C	7967.500	19970.000	-12002.500
1	PAY_W2_C	0.987844	0.214442	3462.0	7173.0	C	8655.000	17932.500	-9277.500
2	PAY_W3_C	0.988290	0.174329	3970.0	6174.0	C	9925.000	15435.000	-5510.000
3	PAY_W4_C	0.975292	0.153386	5766.0	21934.0	C	14415.000	54835.000	-40420.000
4	PAY_W1_S	0.949305	0.153681	18654.0	28699.0	S	9327.000	14349.500	-5022.500
5	PAY_W2_S	0.941286	0.150616	19030.0	37536.0	S	9515.000	18768.000	-9253.000
6	PAY_W3_S	0.932937	0.225260	21537.0	38799.0	S	10768.500	19399.500	-8631.000
7	PAY_W4_S	0.893289	0.134883	735117.0	770874.0	S	367558.500	385437.000	-17878.500
8	PAY_W1_E	0.991191	0.267916	2223.0	4896.0	E	11.115	24.480	-13.365
9	PAY_W2_E	0.990366	0.180820	2713.0	6026.0	E	13.565	30.130	-16.565
10	PAY_W3_E	0.969235	0.175096	5392.0	18429.0	E	26.960	92.145	-65.185
11	PAY_W4_E	0.971831	0.210062	6522.0	18424.0	E	32.610	92.120	-59.510

The DT classifier model achieved an overall subset-hamming score of 96.5% with an overall subset-balanced accuracy of 19.9% (meaning the model prescribes the same actions as the business-as-usual model 19.9% of the time). The model was unable to save costs spent on communication channels (calls, SMS and Emails); instead, the model is costing an extra R108,149.62 on pursuing quarterly debtors to repay their debt. Hence, this model is not cost optimal.

12.4.2 Random forest validation: prescriptive results

The second model fitted is a multilabel RF classifier with parameters set to {bootstrap = True, ccp_alpha, min_impurity_decrease, min_weight_fraction_leaf, random_state, verbose all equal to 0, class_weight, max_depth, max_leaf_nodes max-samples, min_i-mpurity_decrease, n_jobs all equal to None, criterion = gini, max_features = auto, min_samples_leaf = 1, min_samples_split = 2, n_estimators = 100, warm_start = False, base_estimator = RandomForestClassfier and n_jobs = −1} as shown in Figure 12.10.

```
{'estimator__bootstrap': True,
 'estimator__ccp_alpha': 0.0,
 'estimator__class_weight': None,
 'estimator__criterion': 'gini',
 'estimator__max_depth': None,
 'estimator__max_features': 'auto',
 'estimator__max_leaf_nodes': None,
 'estimator__max_samples': None,
 'estimator__min_impurity_decrease': 0.0,
 'estimator__min_impurity_split': None,
 'estimator__min_samples_leaf': 1,
 'estimator__min_samples_split': 2,
 'estimator__min_weight_fraction_leaf': 0.0,
 'estimator__n_estimators': 100,
 'estimator__n_jobs': None,
 'estimator__oob_score': False,
 'estimator__random_state': 0,
 'estimator__verbose': 0,
 'estimator__warm_start': False,
 'estimator': RandomForestClassifier(random_state=0),
 'n_jobs': -1}
```

Figure 12.10: RF model parameters.

The model evaluation was performed using a validation dataset, and the results are summarized in Table 12.2.

The above RF classifier model achieved an overall subset-hamming score of 98.8% with an overall subset-balanced accuracy of 16.82% (meaning the model prescribes similar actions as the business-as-usual model 16.82% of the time). The model was able to save costs spent on communication channels (calls, SMS and Emails) while still retaining repayment behavior from debtors. In total, the model saved R87,904.61, which can be a significant surplus in the long run.

Table 12.2: Random forest evaluation results.

```
Candidate Model: MultiOutputClassifier(estimator=RandomForestClassifier(random_state=0),
                      n_jobs=-1)
[Overall Haming_Score=0.988] [Overall Balanced Accuracy=0.1682]
[Total True_label Action_Costs=R438215.75] [Total Pred_label Action_Costs=R350311.14] ==>Surplus/Deficit=R87904.61
```

	True Columns	Subset_HammingScore	Subset_Bal_Accuracy	True_actions	Pred_actions	string	true_C	pred_C	Savings_R
0	PAY_W1_C	0.996168	0.167532	3187.0	8.0	C	7967.500	20.000	7947.500
1	PAY_W2_C	0.995945	0.201014	3462.0	20.0	C	8655.000	50.000	8605.000
2	PAY_W3_C	0.995388	0.167352	3970.0	27.0	C	9925.000	67.500	9857.500
3	PAY_W4_C	0.993745	0.143589	5766.0	36.0	C	14415.000	90.000	14325.000
4	PAY_W1_S	0.979103	0.143537	18654.0	134.0	S	9327.000	67.000	9260.000
5	PAY_W2_S	0.979038	0.144980	19030.0	103.0	S	9515.000	51.500	9463.500
6	PAY_W3_S	0.973944	0.144943	21537.0	172.0	S	10768.500	86.000	10682.500
7	PAY_W4_S	0.962957	0.116683	735117.0	699757.0	S	367558.500	349878.500	17680.000
8	PAY_W1_E	0.997266	0.250927	2223.0	13.0	E	11.115	0.065	11.050
9	PAY_W2_E	0.996752	0.168115	2713.0	16.0	E	13.565	0.080	13.485
10	PAY_W3_E	0.993477	0.167518	5392.0	38.0	E	26.960	0.190	26.770
11	PAY_W4_E	0.992089	0.202267	6522.0	61.0	E	32.610	0.305	32.305

12.4.3 Multilayer perceptron validation: prescriptive results

The third model fitted is multilabel MLP classifier with parameters set to {activation = relu, alpha = 0.0001, batch_size = auto, beta_1 = 0.9, beta_2 = 0.999, early_stopping = False, epsilon = $1e-08$, hidden_layers_sizer = 100, learning_rate = constant, initialized = 0.001, max_fun = 15,000, max_iter = 200, momentum = 0.9, n_iter_no_change = 10, nesterovs_-momentum = True, power_t = 0.5, shuffle = True, solver = adam, tol = 0.0001, validation_-fraction = 0.1, verbose = False, warm_start = False, base_estimator = MLPClassfier and n_jobs = −1} as shown in Figure 12.11.

The model evaluation was performed using a validation dataset, and the results are summarized in Table 12.3.

The MLP classifier model achieved an overall subset-hamming score of 98.7% with an overall subset-balanced accuracy of 18.1% (meaning the model prescribes similar actions to the business-as-usual model 18.1% of the time). The model was able to save costs spent on communication channels (calls, SMS and Emails) while still retaining repayment behavior from debtors. In total, the model saved R82,005.21, which can be significant in the long run.

12.4.4 Logistic regression validation: prescriptive results

The fourth and last classifier model fitted is multilabel LR with parameters set as {C = 1.0, class_weight = None, dual = False, fit_intercept = True, intercept_scaling = 1, l1_ratios = None, max_iter = 100, multi_class = auto, n_jobs = None, penalty = l2, random_state =

```
{'estimator__activation': 'relu',
 'estimator__alpha': 0.0001,
 'estimator__batch_size': 'auto',
 'estimator__beta_1': 0.9,
 'estimator__beta_2': 0.999,
 'estimator__early_stopping': False,
 'estimator__epsilon': 1e-08,
 'estimator__hidden_layer_sizes': (100,),
 'estimator__learning_rate': 'constant',
 'estimator__learning_rate_init': 0.001,
 'estimator__max_fun': 15000,
 'estimator__max_iter': 200,
 'estimator__momentum': 0.9,
 'estimator__n_iter_no_change': 10,
 'estimator__nesterovs_momentum': True,
 'estimator__power_t': 0.5,
 'estimator__random_state': None,
 'estimator__shuffle': True,
 'estimator__solver': 'adam',
 'estimator__tol': 0.0001,
 'estimator__validation_fraction': 0.1,
 'estimator__verbose': False,
 'estimator__warm_start': False,
 'estimator': MLPClassifier(),
 'n_jobs': -1}
```

Figure 12.11: MLP model parameters.

Table 12.3: Multilayer perceptron evaluation results.

```
Candidate Model: MultiOutputClassifier(estimator=MLPClassifier(), n_jobs=-1)
[Overall Haming_Score=0.9873] [Overall Balanced Accuracy=0.1808]
[Total True_label Action_Costs=R438215.75] [Total Pred_label Action_Costs=R356210.54] ==>Surplus/Deficit=R82005.21
```

	True Columns	Subset_HammingScore	Subset_Bal_Accuracy	True_actions	Pred_actions	string	true_C	pred_C	Savings_R
0	PAY_W1_C	0.995859	0.172466	3187.0	312.0	C	7967.500	780.000	7187.500
1	PAY_W2_C	0.995638	0.207999	3462.0	389.0	C	8655.000	972.500	7682.500
2	PAY_W3_C	0.995278	0.169137	3970.0	183.0	C	9925.000	457.500	9467.500
3	PAY_W4_C	0.993600	0.145467	5766.0	196.0	C	14415.000	490.000	13925.000
4	PAY_W1_S	0.977793	0.147509	18654.0	1379.0	S	9327.000	689.500	8637.500
5	PAY_W2_S	0.977544	0.156988	19030.0	1936.0	S	9515.000	968.000	8547.000
6	PAY_W3_S	0.972533	0.218209	21537.0	1538.0	S	10768.500	769.000	9999.500
7	PAY_W4_S	0.960785	0.118661	735117.0	702151.0	S	367558.500	351075.500	16483.000
8	PAY_W1_E	0.997222	0.252818	2223.0	56.0	E	11.115	0.280	10.835
9	PAY_W2_E	0.996341	0.195182	2713.0	422.0	E	13.565	2.110	11.455
10	PAY_W3_E	0.992704	0.175253	5392.0	808.0	E	26.960	4.040	22.920
11	PAY_W4_E	0.991765	0.210356	6522.0	423.0	E	32.610	2.115	30.495

None, solver = lbfgs, tol = 0.0001, verbose = 0, warm_start = False, base_estimator = Logistic Regression and n_jobs = −1}, as shown in Figure 12.12.

```
{'estimator__C': 1.0,
 'estimator__class_weight': None,
 'estimator__dual': False,
 'estimator__fit_intercept': True,
 'estimator__intercept_scaling': 1,
 'estimator__l1_ratio': None,
 'estimator__max_iter': 100,
 'estimator__multi_class': 'auto',
 'estimator__n_jobs': None,
 'estimator__penalty': 'l2',
 'estimator__random_state': None,
 'estimator__solver': 'lbfgs',
 'estimator__tol': 0.0001,
 'estimator__verbose': 0,
 'estimator__warm_start': False,
 'estimator': LogisticRegression(),
 'n_jobs': -1}
```

Figure 12.12: LR model parameters.

The model evaluation was performed using a validation dataset, and the results are summarized in Table 12.4.

Table 12.4: Logistic regression evaluation results.

```
Candidate Model: MultiOutputClassifier(estimator=LogisticRegression(), n_jobs=-1)
[Overall Haming_Score=0.988] [Overall Balanced Accuracy=0.1631]
[Total True_label Action_Costs=R438215.75] [Total Pred_label Action_Costs=R349711.0] ==>Surplus/Deficit=R88504.75
```

	True Columns	Subset_HammingScore	Subset_Bal_Accuracy	True_actions	Pred_actions	string	true_C	pred_C	Savings_R
0	PAY_W1_C	0.996158	0.166667	3187.0	0.0	C	7967.500	0.0	7967.500
1	PAY_W2_C	0.995944	0.200000	3462.0	0.0	C	8655.000	0.0	8655.000
2	PAY_W3_C	0.995385	0.125000	3970.0	0.0	C	9925.000	0.0	9925.000
3	PAY_W4_C	0.993755	0.142857	5766.0	0.0	C	14415.000	0.0	14415.000
4	PAY_W1_S	0.979146	0.142857	18654.0	0.0	S	9327.000	0.0	9327.000
5	PAY_W2_S	0.979034	0.142857	19030.0	0.0	S	9515.000	0.0	9515.000
6	PAY_W3_S	0.973961	0.142857	21537.0	0.0	S	10768.500	0.0	10768.500
7	PAY_W4_S	0.962977	0.111111	735117.0	699422.0	S	367558.500	349711.0	17847.500
8	PAY_W1_E	0.997265	0.250000	2223.0	0.0	E	11.115	0.0	11.115
9	PAY_W2_E	0.996754	0.166667	2713.0	0.0	E	13.565	0.0	13.565
10	PAY_W3_E	0.993489	0.166667	5392.0	0.0	E	26.960	0.0	26.960
11	PAY_W4_E	0.992079	0.200000	6522.0	0.0	E	32.610	0.0	32.610

The LR classifier model achieved an overall subset-hamming score of 98.8% with an overall subset-balanced accuracy of 16.31% (meaning the model prescribes similar actions to the business-as-usual model 16.31% of the time). The model was able to save costs spent on communication channels (calls, SMS and Emails) while still retaining repayment behavior from debtors. In total, the model saved R88,504.75, which is a significant amount for a business looking to reduce costs and still maximized collections. The only concern with this model is that it only prescribes SMS as a communication method.

12.5 Results

This section will review and compare the trained models to select the best-performing one that is cost-optimal and returns high recovery rates. Since there were four ML models trained, Table 12.5 depicts how they compare against each other.

Table 12.5: Evaluation results of treatment models.

Model	Hamming_Score	Balanced_Accuracy	Action_Cost	Surplus / Deficit
> Logistic Regression	98,8%	16,3%	R349 711,00	R88 504,75
> Random Forest	98,8%	16,8%	R350 311,14	R87 904,61
> Multi-Layer Perceptron	98,7%	18,1%	R356 210,54	R82 005,21
> Decision Tree	96,5%	19,9%	R546 365,38	-R108 149,62

Table 12.5 reveals that the LR and RF classifiers achieved the highest debt recovery rates; however, the LR model only considers SMS as a communication mode, whereas RF is more diverse. For this reason, the RF model is selected as the optimal model to prescribe communication strategies. This finding is based on the highest Hamming score, high subset accuracy, lowest action cost and diversity of communication modes.

12.5.1 Discussion of the outcomes

This section will discuss the key findings of this study, along with the strengths and limitations encountered when building supervised ML models.

12.5.1.1 Key findings

This study proves that to optimize the treatment channel communication strategies, a hybrid of optimization techniques in data preparation and supervised classifier models delivers desirable outcomes.

12.5.1.2 Discussion of key findings

The reviewed literature shows that to model treatment channel prescriptions, ML is an alternative as it is less computationally intensive and does not rely on rules, unlike Markov chains, which need policy constraints to be set. The current study uses a data sample that already has business rules producing optimal treatment channels. The ML model is then used to learn those treatment channels and map them to the relevant debtors in the given sample.

12.5.1.3 Strength and limitations

The application of ML models in this study comes in handy for assigning optimal treatment channels to new datasets without having to retrain. However, the applied ML is only optimal in the assignment of cases and does not optimize the raw data sample.

12.5.1.4 Implications and recommendations

Since the current analysis uses predefined business rules containing treatment channels, it would be worthwhile to train an ML optimization model that would calculate optimal classes for each observation prior to the supervised classification task. The different ML optimization techniques that can be implemented to automatically perform feature engineering include the following:
- Greedy RF tree
- SWARM particle – optimization
- Genetic algorithms
- Heuristics and mathematical programming methods

The above-listed methods should be performed during the feature engineering step, followed by supervised ML modeling as demonstrated in the study.

12.5 Conclusion

The problem this study aims to solve, as supported by [35], reveals that, according to the literature, the Treatment Channel models previously built are not focused on cost optimization but rather on treatment channel prescriptions. For this reason, the research goal was to develop an appropriate solution using supervised ML to optimize the collection process and support a statistical model-building process.

12.5.1 Discussion of research aims

This section discusses each research aim and provides evidence of how the aim was achieved or addressed in the study. The aims that will be addressed were initially mentioned in the first chapter, and the list includes: assessing if the data sample contains erroneous elements, determining if the data sample is normally distributed and does not contain outliers and comparing the test and validation results of the Treatment Channel Model using the relevant evaluation scoring metrics.

12.5.2 Erroneous elements

This research aims to assess the data sample to determine if erroneous elements exist and how they were treated in the study. The erroneous elements encountered included non-value Address type items, inconsistent lengths of postal codes, negative ages of debtors, amounts owed by debtors that were less than R100.00 and negative handover balances.

The non-value Address type field was corrected by mean imputation/replacement; the inconsistent length of the postal code was corrected by applying an algorithmic script that only returned the last four digits. The negative values of the debtor's age and negative handover balance were filtered out, and owed amounts less than R100.00 were also removed by the handover filtering. Hence, the study concludes that the erroneous elements were treated to allow for a smooth and meaningful analysis of the data sample.

12.5.3 Normal distribution and outlier detection

This research aim involves the assessment of the data sample to determine if it comes from a normally distributed population and if it contains outliers. The section concludes by addressing how each point was tackled by the study. The IQR statistic was used to detect evidence of outliers in the sample. The following methods were compared to see which one would scale the data to normality: standard scaler, normalizer, robust scaler, power transformer, quantile transformer, min-max scaler, max absolute scaler and polynomial feature transformation. The Shapiro-Wilk test was then used to confirm if the scaler methods had indeed transformed the data to a state of normality.

The above-mentioned IQR was applied to the sample, and 12 out of 155 variables were detected to have suspected outliers. From the listed scaling methods, the min-max method achieved normality based on the Shapiro-Wilk *p*-value of 100%, which is greater than the 5% alpha value. The same scaling method standardized the outliers and forced them to be between 0.00 and 1.00. Thus, the outliers were kept but controlled in the study. Hence, the study concludes that the outliers were treated, and the sample is now normally distributed, which will allow for a smooth and reliable analysis of the data sample.

12.5.4 Treatment channel model evaluation metrics

This research aims to evaluate ML candidate models used to develop the debtor treatment model/strategy and to present the best candidate model discovered by the study. The candidate ML models compared were all multilabel classifiers, including DT, RF, MLP and LR. The study relied on the following evaluation metrics to identify the best ML model: subset accuracy, Hamming score and collection action cost. The dataset used to evaluate the performance of each model included test and validation datasets.

The ML models were compared, and the best model was a multilabel RF classifier. The best model (RF candidate model) achieved a Hamming score, balanced accuracy and action cost of 98.8%, 16.8% and R350,311.14, respectively. This means that when the normal business strategy is used in recovering debts, it would cost the business, on average, R438,215.75 per month. However, when this model (RF) is used, it would cost, on average, R350,311.14 per month, resulting in a surplus of R87,904.61 in production cost savings. The LR model comes second to the RF model, and its shortfall is in how it only considers the SMS treatment strategy instead of a combination of all actions (SMS, calls and EMAIL). The study concludes that the multilabel RF classifier model is the best model to represent the treatment communication strategy/model, and the same model should be used to prescribe the treatment actions for each debtor to minimize classifier costs, balance communication actions and maximize recovery rates.

In closing, the foremost goal that dictated the objectives of the study was to solve a research problem statement that required a model to optimize banking portfolios. This research goal combines multiple techniques used in the study to determine the best Debt Collection Model. To achieve the best Debt Collection Model, the study proposes the following approach to the problem statement: erroneous elements should always be filtered out; to normalize and treat outliers, MinMax scaler and IQR should be applied; to treat imbalanced datasets, a class_weight hyper parameter should be set to balanced; and to prescribe the final communication actions, a multilabel RF model should be used as it will minimize costs, balance communication actions and maximize recovery rates.

References

[1] C. Romeo and R. Sandler. The effect of debt collection laws on access to credit. Journal of Public Economics, 195:104320, 2021.

[2] L. Mothibi. The Impact of Foreign Debt and Government Debt on Economic Growth in South Africa. In: Proceedings of International Academic Conferences (No. 9912015). International Institute of Social and Economic Sciences, 2019, October.

[3] R. Van de Geer, Q. Wang and S. Bhulai. Data-driven consumer debt collection via machine learning and approximate dynamic programming. Available at SSRN 3250755. 2018.
[4] M. D. O. Dias, A. P. Ribeiro and R. Albergarias. When customers do not pay: A Winning Negotiation Case in Brazil. Journal of Economics and Business, 2(2), 2019.
[5] M. C. So, C. Mues, A. T. De Almeida Filho and L. C. Thomas. Debtor level collection operations using Bayesian dynamic programming. Journal of the Operational Research Society, 70(8):1332–1348, 2019.
[6] L. Phillips and P. Moggridge. Artificial Intelligence in Debt Collection. Credit Control Journal and Asset & Risk Review, 40(2), 2019.
[7] G. Carleo, I. Cirac, K. Cranmer, L. Daudet, M. Schuld, N. Tishby . . . L. Zdeborová. Machine learning and the physical sciences. Reviews of Modern Physics, 91(4):045002, 2019.
[8] G. Wang, K. Wang, Y. Zhou and X. Mo. Establishment of a financial crisis early warning system for domestic listed companies based on three decision tree models. Mathematical Problems in Engineering, 2020:2020.
[9] C. Aroef, Y. Rivan and Z. Rustam. Comparing random forest and support vector machines for breast cancer classification. TELKOMNIKA (Telecommunication Computing Electronics and Control), 18(2), 815–821, 2020.
[10] A. Bellotti, D. Brigo, P. Gambetti and F. Vrins. Forecasting recovery rates on non-performing loans with machine learning. International Journal of Forecasting, 37(1), 428–444, 2021.
[11] K. Potdar, T. S. Pardawala and C. D. Pai. A comparative study of categorical variable encoding techniques for neural network classifiers. International Journal of Computer Applications, 175(4), 7–9, 2017.
[12] A. Bequé and S. Lessmann. Extreme learning machines for credit scoring: An empirical evaluation. Expert Systems with Applications, 86, 42–53. 2017.
[13] A. Shoghi. Debt collection industry: Machine learning approach. Journal of Money and Economy, 14(4), 453–473, 2019.
[14] P. Seragadam, A. Rai, K. C. Ghanta, B. Srinivas, S. K. Lahiri and S. Dutta. Bioremediation of hexavalent chromium from wastewater using bacteria-a green technology. Biodegradation, 32(4):449–466. 2021.
[15] L. A. Drozd and R. Serrano-Padial. Modeling the revolving revolution: The debt collection channel. American Economic Review, 107(3):897–930, 2017.
[16] N. Chehrazi and T. A. Weber. Dynamic valuation of delinquent credit-card accounts. Management Science, 61(12):3077–3096, 2015.
[17] N. Chehrazi, P. W. Glynn and T. A. Weber. Dynamic credit-collections optimization. Management Science, 65(6):2737–2769, 2019.
[18] J. Witzany and A. Kozina. Recovery process optimization using survival regression. Operational Research, 1–28, 2022.
[19] Y. Xue, B. Xue and M. Zhang. Self-adaptive particle swarm optimization for large-scale feature selection in classification. ACM Transactions on Knowledge Discovery from Data (TKDD), 13(5):1–27, 2019.
[20] V. John, A. Boyali, S. Thompson and S. Mita. Bvtnet: Multi-label multi-class fusion of visible and thermal camera for free space and pedestrian segmentation. In: International Conference on Pattern Recognition. Cham: Springer, 277–288, 2021, January.
[21] W. M. Qian and Y. M. Chu. Sharp bounds for a special quasi-arithmetic mean in terms of arithmetic and geometric means with two parameters. Journal of Inequalities and Applications, 2017(1), 1–10, 2017.
[22] P. Mishra, C. M. Pandey, U. Singh, A. Gupta, C. Sahu and A. Keshri. Descriptive statistics and normality tests for statistical data. Annals of cardiac anaesthesia, 22(1):67, 2019.

[23] S. A. Othman and H. T. M. Ali. Improvement of the nonparametric estimation of functional stationary time series using yeo-johnson transformation with application to temperature curves. Advances in Mathematical Physics, 2021:2021.

[24] K. Istenič, N. Gracias, A. Arnaubec, J. Escartín and R. Garcia. Automatic scale estimation of structure from motion based 3D models using laser scalers in underwater scenarios. ISPRS Journal of Photogrammetry and Remote Sensing, 159, 13–25, 2020.

[25] L. Gudmundsson, J. B. Bremnes, J. E. Haugen and T. E. Skaugen. Downscaling RCM precipitation to the station scale using quantile mapping–a comparison of methods. Hydrology & Earth System Sciences Discussions, 9(5), 6185–6201, 2012.

[26] J. H. Heo, H. Ahn, J. Y. Shin, T. R. Kjeldsen and C. Jeong. Probability distributions for a quantile mapping technique for a bias correction of precipitation data: A case study to precipitation data under climate change. Water, 11(7), 1475, 2019.

[27] M. Samnejad, M. Gharib Shirangi and R. Ettehadi. A digital twin of drilling fluids rheology for real-time rig operations. In: Offshore Technology Conference. OnePetro, 2020, May.

[28] S. Calonico, M. D. Cattaneo and M. H. Farrell. Coverage error optimal confidence intervals for local polynomial regression. arXiv Preprint arXiv:1808.01398. 2018.

[29] M. Ekum and A. Ogunsanya. Application of hierarchical polynomial regression models to predict transmission of COVID-19 at global level. International Journal of Clinical Biostatistics and Biometrics, 6(1):27, 2020.

[30] K. Kirasich, T. Smith and B. Sadler. Random forest vs logistic regression: Binary classification for heterogeneous datasets. SMU Data Science Review, 1(3):9, 2018.

[31] A. A. Heidari, H. Faris, I. Aljarah and S. Mirjalili. An efficient hybrid multilayer perceptron neural network with grasshopper optimization. Soft Computing, 23(17):7941–7958, 2019.

[32] S. Dixit, V. Singhal, A. Agarwal and A. P. Rao. Multi-label phase-prediction in high-entropy-alloys using artificial-neural-network. Materials Letters, 268:127606, 2020.

[33] P. Ulmas and I. Liiv. Segmentation of satellite imagery using u-net models for land cover classification. arXiv preprint arXiv:2003.02899. 2020.

[34] B. E. Harcourt. The systems fallacy: A genealogy and critique of public policy and cost-benefit analysis. The Journal of Legal Studies, 47(2):419–447, 2018.

[35] S. C. Onar, B. Oztaysi and C. Kahraman. A fuzzy rule based inference system for early debt collection. Technological and Economic Development of Economy, 24(5):1845–1865, 2018.

[36] P. Bhardwaj. Types of sampling in research. Journal of the Practice of Cardiovascular Sciences, 5(3):157, 2019.

Khadija Rahal*, Zied Driss and Mohamed Salah Abid

Chapter 13
Performance of six turbulence models in predicting two-phase flow on a hydraulic test bench

Abstract: In this study, numerical computations were conducted using the CFD software ANSYS Fluent 17.0. The volume of fluid approach was employed to capture the position of the interface between water and air. The unsteady Reynolds-averaged Navier-Stokes (URANS) equations were used to model the unsteady turbulent flow in a large-scale hydraulic test bench. These equations require turbulence models to close the Reynolds stress tensor. The choice of the turbulence model is a compromise between the accuracy and the computing time, and they can affect the solution accuracy. This study mainly aims to examine six turbulence models and to select the best model that can provide the most accurate simulation of the flow field over the hydraulic test bench. The six turbulence models examined are the standard k-ε model, the realizable k-ε model, the RNG k-ε model, the standard k-ω model, the transition SST k-ω model and the BSL k-ω model. The distribution of the water–air flow characteristics such as the magnitude velocity, the pressure and the turbulence characteristics were presented for each model. The validation of our numerical model was achieved by comparing the present predictions with the experimental data of (Koshizuka et al. 1995) for the breaking of a dam. It is worth noting that the RNG k-ε model provided the best findings along the test section compared to the other models. We can conclude that the RNG k-ε model is the best in terms of accuracy, stability and efficiency of the computed solutions.

Keywords: CFD, turbulence models, URANS modeling, two-phase flow, large-scale hydraulic test bench

*__Corresponding author: Khadija Rahal__, High School of Sciences and Technology of Hammam Sousse (ESSTHS), University of Sousse (US), 4011 Street Lamine Abassi Hammam Sousse, Tunisia; Laboratory of Electro-Mechanic Systems (LASEM), National Engineering School of Sfax (ENIS), University of Sfax (US), B.P. 1173, km 3.5 Soukra, 3038 Sfax, Tunisia, e-mail: khadija.rahal@enis.tn
Zied Driss, Mohamed Salah Abid, Laboratory of Electro-Mechanic Systems (LASEM), National Engineering School of Sfax (ENIS), University of Sfax (US), B.P. 1173, km 3.5 Soukra, 3038 Sfax, Tunisia

https://doi.org/10.1515/9783110717655-013

13.1 Introduction

Turbulence is a physical phenomenon encountered in most flows in nature and in industrial processes. The presence of a wide spectrum of spatial and temporal scales and the strong chaotic nature of the phenomenon are the major characteristics of turbulence [1]. The essence of turbulence is its multi-scale nature. The wide range of spatial and temporal scales inherent in turbulent flows poses interesting challenges to modelers. Darrigol [2] has introduced several theoretical and computational approaches to characterize turbulence, starting from the pioneering works of Prandtl, Kolmogorov and Von Karman. With the growth of computer capabilities, direct numerical simulation (DNS) has been used to investigate various cases of turbulent flows and processes involving the physics of turbulence [3, 4]. Due to the stringent memory storage requirements of DNS, there is a pressing need to reduce the computational cost of simulations. Nevertheless, simplified engineering approximations are becoming increasingly popular and widespread in various industrial engineering applications. Most notably among these, RANS (Reynolds-averaged Navier-Stokes) and LES (large eddy simulation) are the most widely used approaches, although there are several alternatives [5, 6]. Compared to DNS and LES, the RANS approach is computationally inexpensive and can be used to compute any flows encountered in practice. Laurence [7] has provided the insight that advances in computer power have enabled RANS analysis to be applied more broadly to increasingly complex configurations and that, rather than withering away, its use has expanded. This insight has been echoed by [8], who has noted that RANS analysis has become ingrained in engineering practice through increased reliance on user-friendly commercial CFD codes. The RANS approach focuses on the mean flow field using Reynolds decomposition. The main idea of the decomposition is to divide the flow variables into an average and fluctuating quantities. Introducing the Reynolds decomposition into the Navier-Stokes equations leads to an unclosed system of equation. Since, it consists of six more unknown quantities more than equations. The unknowns namely Reynolds stresses, involve the average of the product of fluctuating quantities and need to be modeled for closure. A turbulence model is required to approximate the values of the Reynolds stresses to close the Navier-Stokes equations. Closure models will remain a critical element of computational fluid dynamics analysis into the foreseeable future [9]. Closure modeling remains an active area of research and development. Woelke [10] has discussed the most common two equation eddy viscosity turbulence models generally available in the CFD commercial codes and some of their deficiencies and benefits. These models rely on the turbulent viscosity, which employs the Boussinesq hypothesis for relating the time averaged velocity gradients to the Reynolds stress tensor. In spite of their shortcomings, they present reasonable combination between accuracy and computing time required to predict fluid motion in the complicated configurations. Argyropoulos and Markatos [11] have conducted a detailed review of the turbulence modeling available for predicting turbulent flows, their applications, with a summary of their benefits and drawbacks, and describe several of them. It is used to support future users in selecting the most appropriate turbulence model for their own case of interest. Durbin [12] has surveyed a few selected areas in which closure modeling has advanced in recent years. Duraisamy et al. [13] have reviewed the current developments in bounding uncertainties in RANS models via physical constraints, in adopting statistical inference to characterize model coefficients and estimate discrepancy, and in using machine learning technique to improve turbulence models. Turbulence closure modeling has been the subject of many other reviews [14–18], and studies [19–22], and is commonly reviewed in books on CFD and manuals on CFD software. The present work aims to conduct a detailed investigation to evaluate the performance of six turbulence models regarding their accuracy in modeling a complex multiphase flow field in a hydraulic test bench and to determine in particular which turbulence model produces the best numerical findings.

13.2 Physical facility

The large-scale hydraulic test bench is depicted in Figure 13.1. The water runs in a closed circuit composed of a tank, a penstock, a test section and a sluice gate. The tank has a rectangular shape with a length of 1.14 m, a width of 0.9 m and a height of 1 m. The penstock has a rectangular shape with a width of 0.4 m and a length of 1.72 m, connected to a test section with a slope of 45°, made of Plexiglas plates. The test section is about 3 m in length, 0.4 m in height and 0.4 m in width.

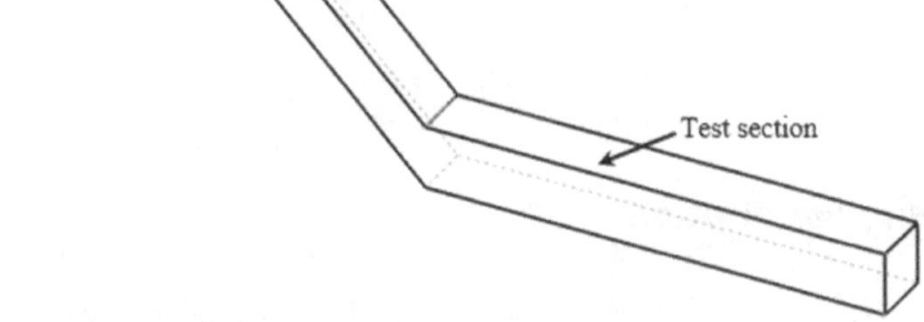

Figure 13.1: Physical facility.

13.3 Numerical model

13.3.1 URANS governing equation

In this investigation, the CFD software ANSYS Fluent 17.0 was used to perform the 3D modeling. To simulate the two-phase flow over hydraulic test bench, the unsteady Reynolds-averaged Navier-Stokes (URANS) equations are employed. The continuity and momentum equations are given by the following set of equations [23]:

$$\frac{\partial \rho}{\partial t} + \frac{\partial \rho u_i}{\partial x_i} = 0 \tag{13.1}$$

$$\frac{\partial \rho u_i}{\partial t} + \frac{\partial}{\partial x_j}\left(\rho u_i u_j\right) = -\frac{\partial P}{\partial x_i} + \frac{\partial}{\partial x_j}\left\{\mu\left(\frac{\partial u_i}{\partial x_j} + \frac{\partial u_j}{\partial x_i}\right)\right\} - \frac{\partial \rho \overline{u'_i u'_j}}{\partial x_j} + \rho g_i \tag{13.2}$$

where u_i and u_j are the mean velocities, x_i and x_j are the components of the Cartesian coordinate system ($i, j = 1, 2, 3$)), P denotes the pressure, ρ is the density and μ refers to the dynamic viscosity. Additionally, g_i represents the gravitational acceleration, and t is the time.

The sophistication of the RANS equation is due to the appearance of a new turbulent term, referred to as the Reynolds stress. $\tau_{ij} = -\rho\overline{u'_i u'_j}$. The physical significance of this term is the average effect of turbulent advection on the mean flow [24]. With the inclusion of this new term, the URANS equation is unclosed, with an additional six unknowns appearing from the Reynolds stress tensor. The general URANS equations have a total of ten unknowns (the mean pressure, the three mean velocity components and the six independent components of the Reynolds stress tensor). In order to tackle the problem, enough equations must be found to solve all the unknowns. The nine components of the Reynolds stress tensor can be expressed as follows:

$$\tau_{ij} = -\rho\overline{u'_i u'_j} = -\rho\begin{bmatrix} \overline{u'_1 u'_1} & \overline{u'_1 u'_2} & \overline{u'_1 u'_3} \\ \overline{u'_2 u'_1} & \overline{u'_2 u'_2} & \overline{u'_2 u'_3} \\ \overline{u'_3 u'_1} & \overline{u'_3 u'_2} & \overline{u'_3 u'_3} \end{bmatrix} \tag{13.3}$$

13.3.2 Eddy viscosity assumption

The concept of eddy viscosity is based on the hypothesis that the mean-flow velocity gradients can be related to the turbulent stresses to close the system of equations. This hypothesis is recognized as the Boussinesq assumption, and it gives the following expression for the Reynolds stress:

$$\tau_{ij} = \mu_t \left(\frac{\partial U_i}{\partial x_j} + \frac{\partial U_j}{\partial x_i} \right) - \frac{2}{3} \delta_{ij} \left(\rho k + \mu_t \frac{\partial U_k}{\partial x_k} \right) \tag{13.4}$$

which can be written in shorthand as

$$\tau_{ij} = 2\mu_t S_{ij} - \frac{2}{3} \rho k \delta_{ij} \tag{13.5}$$

where S_{ij} denotes the mean rate of strain tensor, μ_t denotes the turbulence eddy viscosity, k is the turbulent kinetic energy, δ_{ij} is the Kronecker delta and τ_{ij} stands for the viscous tensor.

Another basic concept in the conceptualization of closure models is that of turbulent kinetic energy. It represents the amount of energy stored in turbulent fluctuations. The turbulent kinetic energy is expressed as

$$k = \frac{1}{2} \overline{u_i' u_i'} = \frac{1}{2} \left(\overline{u_1'^2} + \overline{u_2'^2} + \overline{u_3'^2} \right) \tag{13.6}$$

13.3.3 Turbulence models

One of the key elements in the success of most CFD simulations is turbulence modeling. It is worth mentioning that choosing a suitable turbulence model is a difficult of numerical simulation in hydraulic. In particular, it is inevitably an exercise of compromise between computational cost and predictive accuracy. Therefore, we should compare the different turbulence models and select the most appropriate one for the relevant problem. The ANSYS FLUENT code has several turbulent models for turbulent viscosity, such as the standard *k*-*ε* model, the RNG *k*-*ε* model, the realizable *k*-*ε* model, the *k*-*ω* model, the shear stress transport (SST) model and the BSL *k*-*ω* model.

13.3.3.1 Standard *k*-*ε* model

The standard *k*-*ε* model was initially introduced by [24]. The *k*-*ε* model includes the effect of molecular viscosity in the regions near the wall. The equations for the turbulent kinetic energy k and its rate of dissipation ε are defined as follows:

$$\frac{\partial}{\partial t}(\rho k) + \frac{\partial}{\partial x_i}(\rho k u_i) = \frac{\partial}{\partial x_j} \left[\left(\mu + \frac{\mu_t}{\sigma_k} \right) \frac{\partial k}{\partial x_j} \right] - \rho \varepsilon + \mu_t S^2 \tag{13.7}$$

$$\frac{\partial}{\partial t}(\rho \varepsilon) + \frac{\partial}{\partial x_i}(\rho \varepsilon u_i) = \frac{\partial}{\partial x_j} \left[\left(\mu + \frac{\mu_t}{\sigma_\varepsilon} \right) \frac{\partial \varepsilon}{\partial x_j} \right] + C_{1\varepsilon} \frac{\varepsilon}{k} \mu_t S^2 - C_{2\varepsilon} \rho \frac{\varepsilon^2}{k} \tag{13.8}$$

The turbulent viscosity μ_t is computed as follows:

$$\mu_t = \rho C_\mu \frac{k^2}{\varepsilon} \tag{13.9}$$

where S is the mean rate-of-strain tensor, it is defined as

$$S \equiv \sqrt{2S_{ij}S_{ij}} \tag{13.10}$$

with the mean strain rate S_{ij} provided by

$$S_{ij} = \frac{1}{2}\left(\frac{\partial u_i}{\partial x_j} + \frac{\partial u_j}{\partial x_i}\right) \tag{13.11}$$

The default values of the related parameters are

$$C_\mu = 0.09,\ C_{1\varepsilon} = 1.44,\ C_{2\varepsilon} = 1.92,\ \sigma_k = 1.0 \text{ and } \sigma_\varepsilon = 1.3.$$

13.3.3.2 Realizable *k-ε* model

Shih et al. [25] proposed the realizable k-ε model. This model is novel and shows an advancement over the standard k-ε model, satisfying a number of mathematical limitations on the Reynolds stress. The turbulent kinetic energy k and its rate of dissipation ε are presented in the following equations:

$$\frac{\partial}{\partial t}(\rho k) + \frac{\partial}{\partial x_j}(\rho k u_j) = \frac{\partial}{\partial x_j}\left[\left(\mu + \frac{\mu_t}{\sigma_k}\right)\frac{\partial k}{\partial x_j}\right] - \rho\varepsilon + \mu_t S^2 \tag{13.12}$$

$$\frac{\partial}{\partial t}(\rho \varepsilon) + \frac{\partial}{\partial x_j}(\rho \varepsilon u_j) = \frac{\partial}{\partial x_j}\left[\left(\mu + \frac{\mu_t}{\sigma_\varepsilon}\right)\frac{\partial \varepsilon}{\partial x_j}\right] + \rho C_1 S\varepsilon - C_2 \rho \frac{\varepsilon^2}{k + \sqrt{\vartheta\varepsilon}} \tag{13.13}$$

where

$$C_1 = \max\left[0.43, \frac{\eta}{\eta + 5}\right], \quad \eta = \sqrt{2S_{ij}S_{ij}}\frac{k}{\varepsilon} \tag{13.14}$$

The turbulent viscosity μ_t is computed according to eq. (13.9):

$$\mu_t = \rho C_\mu \frac{k^2}{\varepsilon}$$

The default values of related parameters are

$$C_{1\varepsilon} = 1.44,\ C_2 = 1.9,\ \sigma_k = 1.0 \text{ and } \sigma_\varepsilon = 1.2$$

13.3.3.3 RNG *k-ε* model

The RNG *k-ε* model was rooted in the ideas of Yakhot and Orszag [26]. This model originated from using a mathematical technique known as the "renormalization group" (RNG) method from the instantaneous Navier-Stokes equations. The RNG *k-ε* model is different from the standard *k-ε* model in terms of the model constants, additional terms and functions. The equations for the turbulent kinetic energy k and its rate of dissipation ε are defined as follows:

$$\frac{\partial}{\partial t}(\rho k) + \frac{\partial}{\partial x_i}(\rho k u_i) = \frac{\partial}{\partial x_j}\left[\alpha_k \mu_{eff} \frac{\partial k}{\partial x_j}\right] - \rho\varepsilon + \mu_t S^2 \tag{13.15}$$

$$\frac{\partial}{\partial t}(\rho \varepsilon) + \frac{\partial}{\partial x_i}(\rho \varepsilon u_i) = \frac{\partial}{\partial x_j}\left[\alpha_\varepsilon \mu_{eff} \frac{\partial \varepsilon}{\partial x_j}\right] + C_{1\varepsilon}\frac{\varepsilon}{k}\mu_t S^2 - C_{2\varepsilon}\rho\frac{\varepsilon^2}{k} - R_\varepsilon \tag{13.16}$$

The turbulent viscosity μ_t is computed according to eq. (13.9):

$$\mu_t = \rho C_\mu \frac{k^2}{\varepsilon}$$

The additional term given in the following equation clearly shows the distinction between the standard *k-ε* and RNG models:

$$R_\varepsilon = \frac{C_\mu \rho \eta^3 (1 - \eta/\eta_0)}{1 + \beta\eta^3}\frac{\varepsilon^2}{k} \tag{13.17}$$

where

$$\eta \equiv Sk/\varepsilon$$

where η is given by eq. (13.14), $\eta_0 = 4.38$ and $\beta = 0.012$.

Based on eq. (13.17), the third and fourth terms on the right-hand side of eq. (13.16) can be merged, and the resulting ε equation can be rewritten as

$$\frac{\partial}{\partial t}(\rho \varepsilon) + \frac{\partial}{\partial x_i}(\rho \varepsilon u_i) = \frac{\partial}{\partial x_j}\left[\alpha_\varepsilon \mu_{\text{eff}} \frac{\partial \varepsilon}{\partial x_j}\right] + C_{1\varepsilon}\frac{\varepsilon}{k}\mu_t S^2 - C_{2\varepsilon}^*\rho\frac{\varepsilon^2}{k} \tag{13.18}$$

where $C_{2\varepsilon}^*$ is provided by

$$C_{2\varepsilon}^* \equiv C_{2\varepsilon} + \frac{C_\mu \eta^3 (1 - \eta/\eta_0)}{1 + \beta\eta^3} \tag{13.19}$$

The default values of related parameters are

$$C_{1\varepsilon} = 1.42,\ C_{2\varepsilon} = 1.68,\ C_{2\varepsilon}^* = 2.0$$

13.3.3.4 Standard *k-ω* model

The standard k-ω model is based on the Wilcox k-ω model [27]. The equations for the turbulent kinetic energy k and its specific dissipation rate ω are defined as follows:

$$\frac{\partial}{\partial t}(\rho k) + \frac{\partial}{\partial x_j}(\rho k u_j) = \frac{\partial}{\partial x_j}\left[(\mu + 0.5\mu_t)\frac{\partial k}{\partial x_j}\right] + \mu_t S^2 - \rho\beta^* \omega k \tag{13.20}$$

$$\frac{\partial}{\partial t}(\rho \omega) + \frac{\partial}{\partial x_j}(\rho \omega u_j) = \frac{\partial}{\partial x_j}\left[(\mu + 0.5\mu_t)\frac{\partial \omega}{\partial x_j}\right] + \left(\frac{\alpha\omega}{k}\right)\mu_t S^2 - \rho\beta\omega^2 \tag{13.21}$$

The turbulent viscosity μ_t is computed by combining k and ω as follows:

$$\mu_t = \alpha^* \frac{\rho k}{\omega} \tag{13.22}$$

where ω is the specific dissipation rate.

The default values of the related parameters are

$$\alpha^* = 1,\ \alpha = 0.52,\ \beta^* = 9/100,\ \beta = 0.0072$$

13.3.3.5 Shear stress transport *k-ω* model (SST)

Menter [28] developed the shear stress transport (SST) k-ω model, which calculates the turbulent flow field using the standard k-ω model in the near-wall region and switches to the standard k-ε model in the far field. The equations for the turbulence kinetic energy k and its specific dissipation rate ω are defined as follows:

$$\frac{\partial}{\partial t}(\rho k) + \frac{\partial}{\partial x_j}(\rho k u_j) = \frac{\partial}{\partial x_j}\left[\left(\mu + \frac{\mu_t}{\sigma_k}\right)\frac{\partial k}{\partial x_j}\right] + \tilde{G}_k - \rho\beta^* \omega k \tag{13.23}$$

$$\frac{\partial}{\partial t}(\rho \omega) + \frac{\partial}{\partial x_j}(\rho \omega u_j) = \frac{\partial}{\partial x_j}\left[\left(\mu + \frac{\mu_t}{\sigma_\omega}\right)\frac{\partial \omega}{\partial x_j}\right] + \left(\frac{\alpha\omega}{k}\right)\mu_t S^2 - \rho\beta\omega^2 + 2(1 - F_1)\rho\sigma_{\omega,2}\frac{1}{\omega}\frac{\partial k}{\partial x_j}\frac{\partial \omega}{\partial x_j} \tag{13.24}$$

The turbulent viscosity μ_t is computed as follows:

$$\mu_t = \frac{\rho k}{\omega}\frac{1}{\max\left[\frac{1}{\alpha^*}, \frac{SF_2}{a_1\omega}\right]} \tag{13.25}$$

where

$$\tilde{G}_k = \min\left(\mu_t S^2, 10\rho\beta^* k\omega\right) \tag{13.26}$$

The default values of the related parameters are

$$\alpha_1 = 0.31, \alpha^* = 1, \alpha = 0.52, \beta^* = 9/100 \text{ and } \beta = 0.0072$$

13.3.3.6 Baseline (BSL) *k-ω* model

To reduce the limitations of the k-ω model by Wilcox, which is strongly sensitive to free stream conditions, Menter [28] developed the baseline (BSL) k-ω model. This model is based on a combination of the k-ω model in the inner region of the boundary layer and the k-ε model in the outer region and in the free shear layers. The transport equations for the baseline model are expressed in the following form:

$$\frac{\partial}{\partial t}(\rho k) + \frac{\partial}{\partial x_j}(\rho k u_j) = \frac{\partial}{\partial x_j}\left[\left(\mu + \frac{\mu_t}{\sigma_k}\right)\frac{\partial k}{\partial x_j}\right] + \mu_t S^2 - \rho\beta^*\omega k \tag{13.27}$$

$$\frac{\partial}{\partial t}(\rho\omega) + \frac{\partial}{\partial x_j}(\rho\omega u_j) = \frac{\partial}{\partial x_j}\left[\left(\mu + \frac{\mu_t}{\sigma_\omega}\right)\frac{\partial \omega}{\partial x_j}\right] + \left(\frac{\alpha\omega}{k}\right)\mu_t S^2 - \rho\beta\omega^2 + 2\rho\frac{1}{\sigma_{\omega,2}}\frac{1}{\omega}\frac{\partial k}{\partial x_j}\frac{\partial \omega}{\partial x_j} \tag{13.28}$$

The addition of a new cross-diffusion term is the major distinction between this formulation and the original k-ω model in the ω-equation, and the constants of the model are different [29]. Furthermore, the standard k-ω model was multiplied by a blending function F_1. To create the new model, the transformed equations were multiplied by $(1–F_1)$, and the corresponding equations of each set were added together [29]:

$$\frac{\partial}{\partial t}(\rho k) + \frac{\partial}{\partial x_j}(\rho k u_j) = \frac{\partial}{\partial x_j}\left[\left(\mu + \frac{\mu_t}{\sigma_k}\right)\frac{\partial k}{\partial x_j}\right] + \mu_t S^2 - \rho\beta^*\omega k \tag{13.29}$$

$$\frac{\partial}{\partial t}(\rho\omega) + \frac{\partial}{\partial x_j}(\rho\omega u_j) = \frac{\partial}{\partial x_j}\left[\left(\mu + \frac{\mu_t}{\sigma_\omega}\right)\frac{\partial \omega}{\partial x_j}\right]$$
$$+ \left(\frac{\alpha\omega}{k}\right)\mu_t S^2 - \rho\beta\omega^2 + 2\rho(1 - F_1)\frac{1}{\sigma_{\omega,2}}\frac{1}{\omega}\frac{\partial k}{\partial x_j}\frac{\partial \omega}{\partial x_j} \tag{13.30}$$

The blending function F_1 is given by

$$F_1 = \tanh\left(\Phi_1^4\right) \tag{13.31}$$

with

$$\Phi_1 = \min\left[\max\left(\frac{\sqrt{k}}{0.09\omega y}, \frac{500\mu}{\rho y^2\omega}\right), \frac{4\rho k}{\sigma_{\omega,2}D_\omega^+ y^2}\right] \tag{13.32}$$

$$D_\omega^+ = \max\left[2\rho\frac{1}{\sigma_{\omega,2}}\frac{1}{\omega}\frac{\partial k}{\partial x_j}\frac{\partial \omega}{\partial x_j}, 10^{-10}\right] \tag{13.33}$$

where y refers to the distance to the next surface and D_ω^+ is the positive portion of the cross-diffusion term.

The BSL k-ω model is based on both the standard k-ε and the standard k-ω models. The standard k-ε model has been transformed into equations based on k and ω to blend the robust and accurate formulations of the two models. This leads to the introduction of a cross-diffusion term, D_ω.

D_ϖ is defined as

$$D_\omega = 2(1-F_1)\rho\frac{1}{\omega\sigma_{\omega,2}}\frac{\partial k}{\partial x_j}\frac{\partial \omega}{\partial x_j} \tag{13.34}$$

The default values of the related parameters are

$$\sigma_{k,1} = 2.0,\ \sigma_{\omega,1} = 2.0,\ \sigma_{k,2} = 1.0,\ \sigma_{\omega,2} = 1.168,\ \beta_{i,1} = 0.075,\ \beta_{i,2} = 0.0828$$

3.4 Free surface modeling

The volume of fluid (VOF) method was suggested by [30]. It is based on the concept of a volume fraction and the fact that the phases are not interpenetrating. This method is one of the most widely used approaches in various commercial CFD software to capture the free surface position. The interface between the water and air was reached through resolving the advection equation for each phase throughout the domain:

$$\frac{\partial \alpha}{\partial t} + u_i\frac{\partial \alpha}{\partial x_i} = 0 \tag{13.35}$$

where u_i denotes the velocity of the flow and x_i the coordinate in direction i ($i = 1, 2$ or 3) and t represents time.

The VOF method, based on the field volume fraction denoted as α, can have the following values:

$$\begin{cases} \alpha = 0\text{: the cell is empty} \\ \alpha = 1\text{: the cell is full} \\ 0 < \alpha < 1\text{: the cell contains the surface} \end{cases}$$

In a two-phase system, the volume fraction of the liquid is tracked and the mixture properties of the air and water phases, based on the volume fraction weighted average, are used [29]. The density and viscosity in each cell at the interface are presented in the following equations:

$$\rho = \rho_l \alpha_l + (1 - \alpha_l)\rho_g \tag{13.36}$$

$$\mu = \mu_l \alpha_l + (1 - \alpha_l)\mu_g \tag{13.37}$$

where α is the volume fraction, ρ_g, ρ_l are the gas and liquid densities, and μ_l, μ_g are the gas and liquid viscosities.

13.4 Boundary conditions

Determining boundary conditions is considered one of the most crucial factors that can influence the accuracy of the results. Consequently, a precise representation of the boundary conditions may be more significant than the selection of the mesh and the turbulence model. The appropriate boundary conditions used in the numerical analyses are also indicated in Figure 13.2. The boundary at the upstream of the computational domain was adjusted as a velocity inlet. An outlet boundary condition was applied to the downstream of the computational domain, and the pressure at the outlet was set to 0 Pa gauge pressure. The no-slip condition is assumed to hold at the walls, which states that the velocity at the wall must be equal to zero ($u = v = w = 0$). The acceleration of gravity was applied in the negative y-direction.

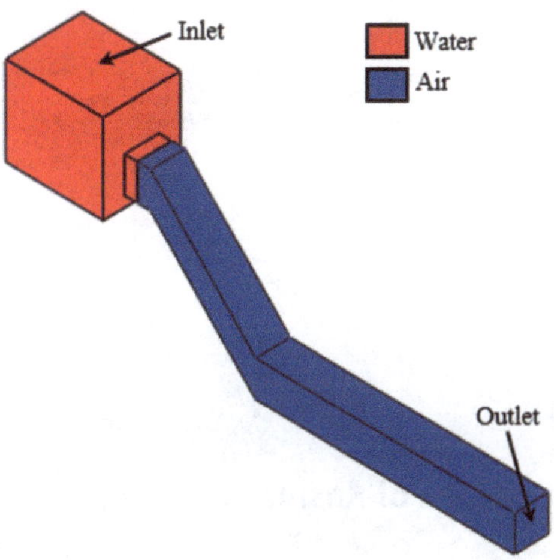

Figure 13.2: Boundary Conditions.

13.5 Computational meshing

The mesh is arguably one of the most important keys in a good CFD simulation, which requires much attention. The mesh can tremendously influence the accuracy of the results, its turnaround time and hence its cost. If we find non-orthogonal elements in the mesh, the VOF model is susceptible to instabilities. Therefore, we should note that the mesh used for the multiphase simulations must be carefully selected to ensure for acceptable accuracy and numerical stability. Increasing the density of the mesh may improve the accuracy of simulation results. Nevertheless, a higher grid density will require more computing time and increased CPU utilization. The mesh in the computational domain was constructed by employing ANSYS Meshing software. The created final computational mesh has 544,187 elements, and it was generated in a structured grid to save computing power, to improve convergence rate and produce more accurate results. The computational mesh is presented in Figure 13.3. The mesh quality plays a great role in the stability and accuracy of the numerical simulation.

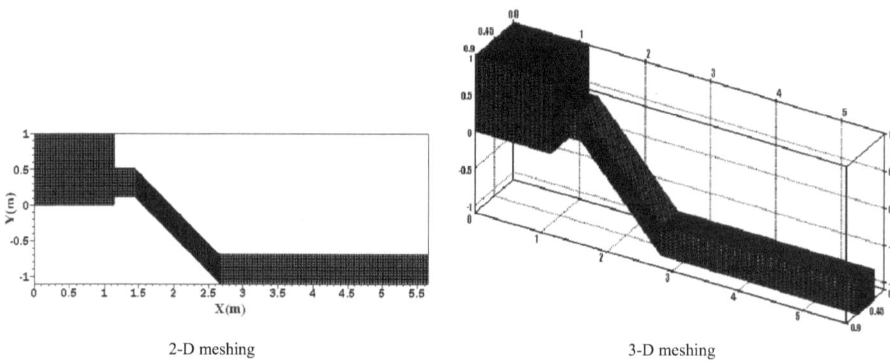

Figure 13.3: Meshing of the computational domain.

13.6 Results and discussion

13.6.1 Comparison with laboratory data of Koshizuka et al.,

Validating numerical simulations is often a considerable challenge, as these models rely on numerous assumptions that can significantly impact the outcomes. These models should be verified to ensure the validity of our numerical methods. Validation is performed by comparing the findings of the numerical simulations for the normalized evolution of the water level H^* versus time τ^* achieved with the ANSYS Fluent software, with the results of measurements from [31]. Figure 13.4 displays the comparison between the CFD results and experiments of [31]. The numerical results match very well the experimental findings.

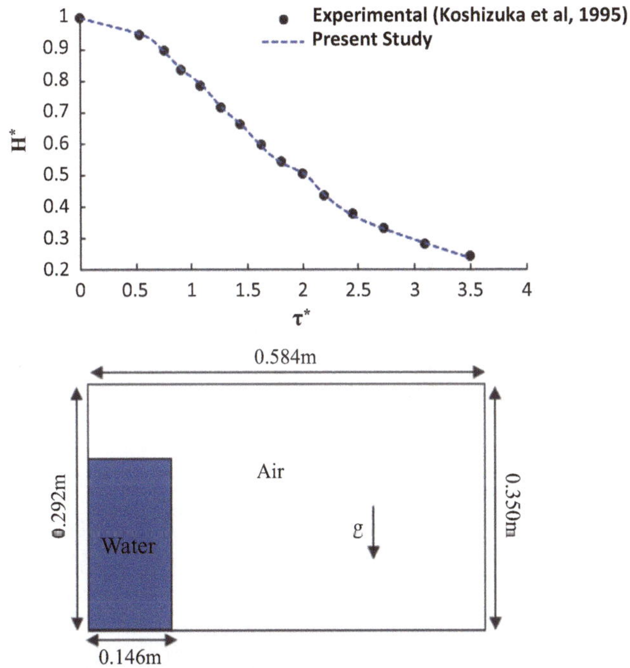

Figure 13.4: Numerical model validation against the experimental data of Koshizuka et al., (a) Normalized evolution of the water level H* versus time τ*. (b) 2D sketch of the configuration.

13.6.2 Water volume fraction

Figure 13.5 depicts the volume fraction distribution of water in the plane defined by $z = 0.45$ m and $x = 3.6$ m over the hydraulic test bench, as predicted by the standard k-ε model, the realizable k-ε model, the RNG k-ε model, the standard k-ω model, the transition SST model and the BSL k-ω model. The VOF method separates both air and water with high accuracy. Overall, we can observe that similar findings are obtained with all turbulence models. For all six cases, the computed free surface acquires an expected and acceptable shape. The volume of an air-phase cell is zero; a value of one indicates that the cell is full of water, while $0 < C < 1$ indicates that the cell comprises the free surface as it is partially filled with water. It is remarkable that the structure of air–water flow is clearly dominated by collisions and interactions between particles with the flow turbulence. These processes lead to continuous breakup and interfacial deformations that pave the way for large air–water interfacial velocities. Continuous changes in air–water interfaces may lead to high levels of turbulence. As seen from these results, the calculation was started by removing the gate, and the upstream water jet fell down in the stream-wise direction. In addition, it has been observed that the water accelerates significantly due to the effect of gravity along the waterfall. This zone is characterized by a significant increase in pressure, leading to a strong pressure gradient pushing the water particles downstream.

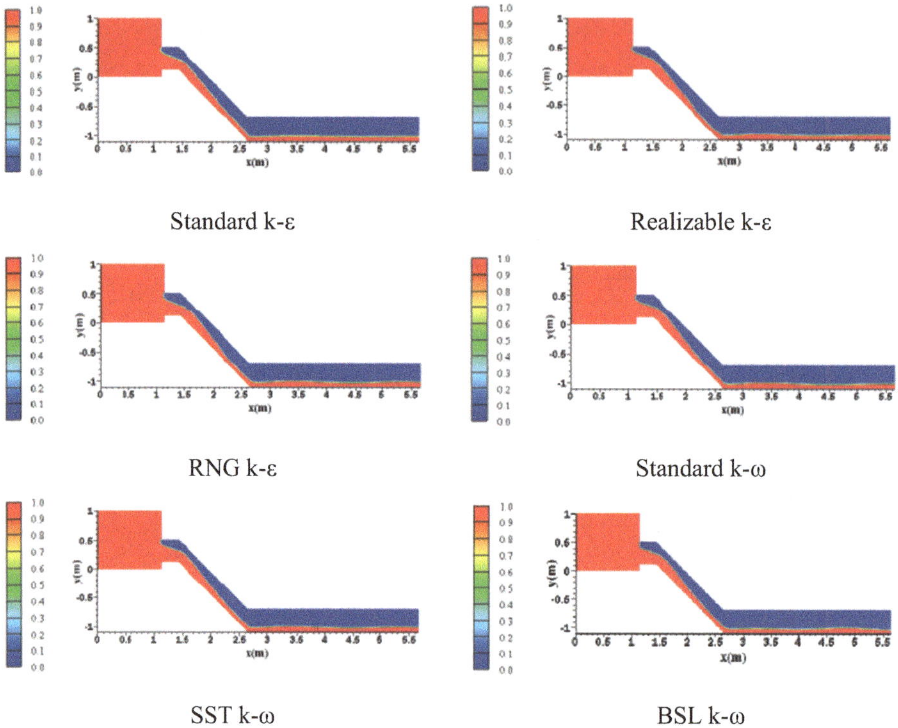

Figure 13.5: Volume fraction distribution.

13.6.3 Magnitude velocity

Figure 13.6 illustrates the velocity distribution of the water flow in the plane defined by $z = 0.45$ m and $x = 3.6$ m over the hydraulic test bench, as predicted by the standard k-ε model, the realizable k-ε model, the RNG k-ε model, the standard k-ω model, the transition SST model and the BSL k-ω model. It is obvious that the six models tended to give almost the same results, beginning with velocity gradually increasing from the penstock until a maximum velocity was reached. Consequently, the water flow is accelerated by the gravity effect along the hydraulic test bench. The velocity increases and develops in the downstream direction, becoming relatively constant in the test section. Therefore, higher velocities appear through the test section and can cause more turbulence. For example, we can see that the maximum value of the magnitude of water velocity is equal to $V = 5.28$ m/s with the standard k-ε model, $V = 5.59$ m/s with the realizable k-ε model, $V = 5.61$ m/s with the RNG k-ε model, $V = 5.55$ m/s with the standard k-ω model, $V = 5.60$ m/s with the transition-SST model and $V = 5.48$ m/s with the BSL SST k-ω model. However, near the free surface, the velocities of water and air are approximately equal, but the velocities of the air decrease to zero in the region closest to the wall. The comparison between these results shows that the turbulence models directly affect the magnitude of water velocity in the hydraulic test bench.

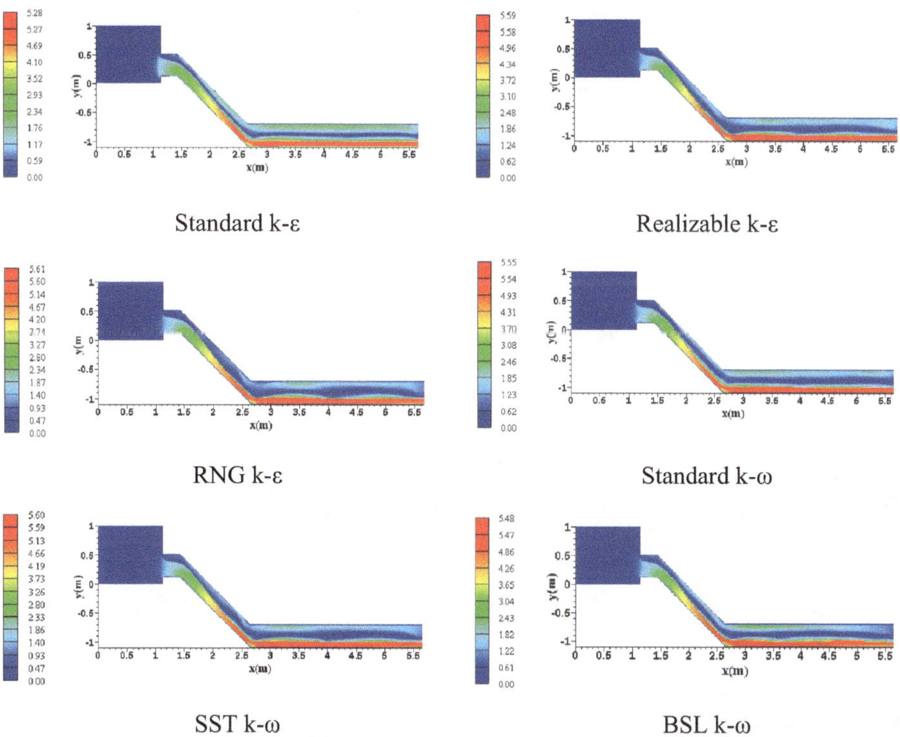

Figure 13.6: Magnitude velocity distribution.

13.6.4 Static pressure

Figure 13.7 presents the static pressure distribution in the plane defined by $z = 0.45$ m and $x = 3.6$ m over the hydraulic test bench, predicted by the standard k-ε model, the realizable k-ε model, the RNG k-ε model, the standard k-ω model, the transition SST model and the BSL k-ω model. It can be observed that the static pressure tends to yield similar results. The maximum value of the static pressure is caused by the impact of the falling water from the tank directly hitting the penstock and the test section. The maximum pressure value is equal to $p = 10{,}800.57$ Pa with the standard k-ε model, $p = 10{,}462.92$ Pa with the realizable k-ε model, $p = 10{,}150.27$ Pa with the RNG k-ε model, $p = 10{,}520.63$ Pa with the standard k-ω model, $p = 10169.94$ Pa with the transition-SST model and $p = 10420.71$ Pa with the BSL k-ω model. Additionally, the minimum pressure was obtained at the top of the tank and then increased to reach a maximum in the penstock and the test section. Based on the simulated findings, it was found that there is negative pressure at the top. It can also be noted that the maximum negative pressure equals $p = -4{,}753.63$ Pa with the standard k-ε model, $p = -4{,}753.66$ Pa with the realizable k-ε model, $p = -4{,}748.50$ Pa with the RNG k-ε model, $p = -4{,}764.63$ Pa with the standard k-ω model, $p = -4{,}747.69$ Pa with the transition-SST model and $p = -4{,}747.83$ Pa with the BSL k-ω model. The static pressure has a significant influence on the fluid flow. The negative pressure plays a significant role in the evaluation of cavitation potential. The pressure distribution is not uniform and shows an undulating pattern along the length of the hydraulic test bench.

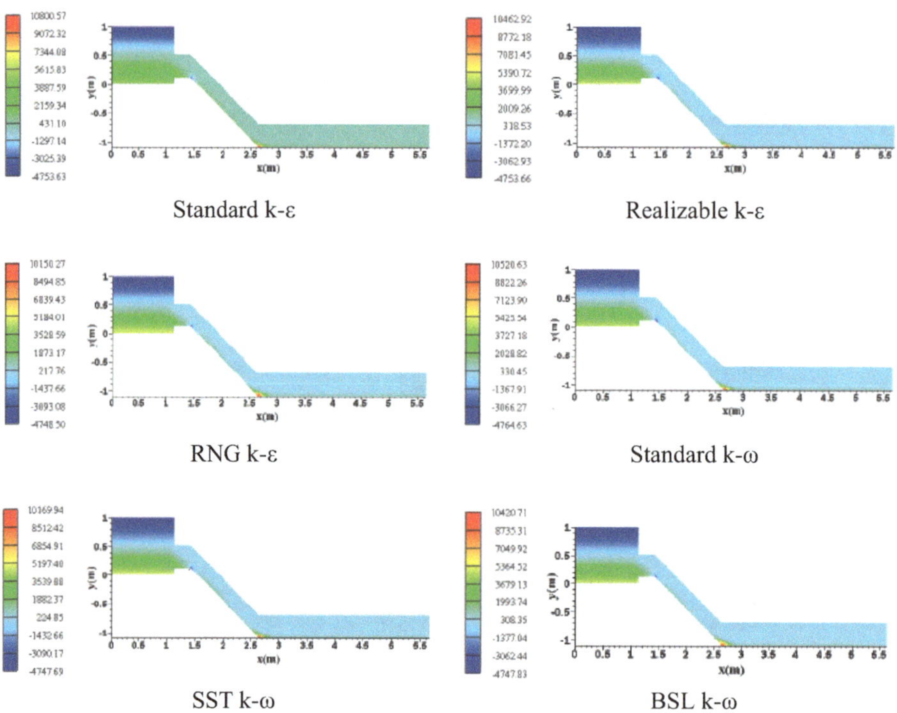

Figure 13.7: Static pressure distribution.

13.6.5 Dynamic pressure

Dynamic pressure arises from the motion of the fluid and is closely related to the kinetic energy of its particles. Figure 13.8 describes the dynamic pressure distribution in the plane defined by $z = 0.45$ m and $x = 3.6$ m over the hydraulic test bench, as predicted by the standard *k-ε* model, the realizable *k-ε* model, the RNG *k-ε* model, the standard *k-ω* model, the transition SST model and the BSL *k-ω* model. It can be observed that the distribution of dynamic pressure is nearly the same for the different turbulence models. The highest pressure is observed at the penstock and the test section, produced by the conversion of the gravitational potential energy of the moving water into kinetic energy. The maximum pressure value is equal to $p = 13{,}926.58$ Pa with the standard *k-ε* model, $p = 14{,}653.95$ Pa with the realizable *k-ε* model, $p = 14{,}925.70$ Pa with the RNG *k-ε* model, $p = 14{,}449.84$ Pa with the standard *k-ω* model, $p = 14{,}880.49$ Pa with the transition-SST model and $p = 14{,}492.97$ Pa with the BSL *k-ω* model. Dynamic pressure distribution depends on the velocity of the fluid, the density of the fluid and the acceleration due to gravity. This pressure becomes higher when the medium is dense, like water.

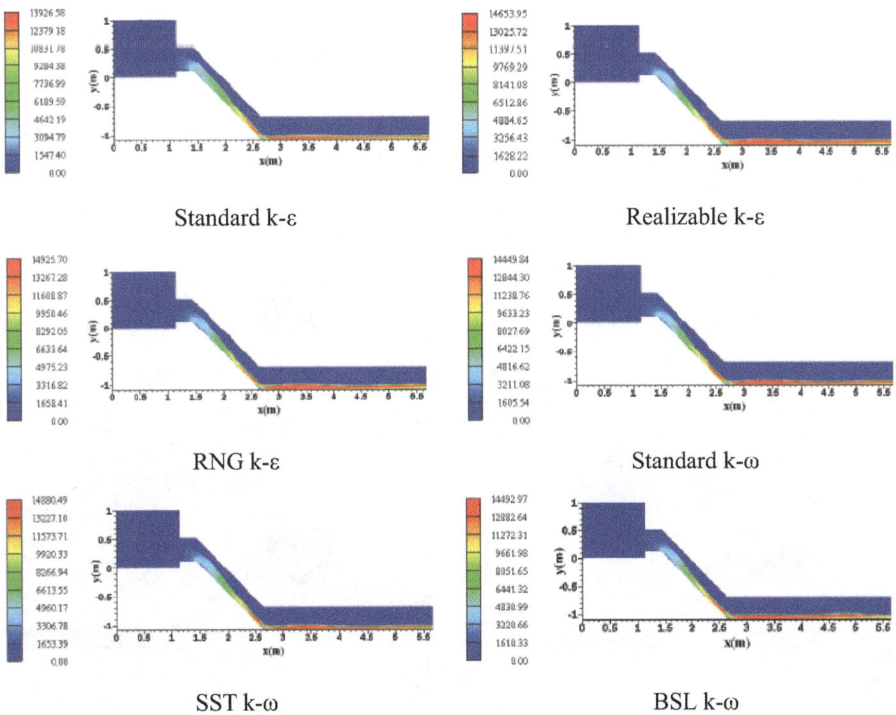

Figure 13.8: Dynamic pressure distribution.

13.6.6 Turbulent kinetic energy

The distribution of the turbulent kinetic energy in the plane defined by $z = 0.45$ m and $x = 3.6$ m over the hydraulic test bench, predicted by the standard k-ε model, the realizable k-ε model, the RNG k-ε model, the standard k-ω model, the transition SST model and the BSL k-ω model, is depicted in Figure 13.9. We can observe that the turbulent kinetic energy differs among turbulence models. From top to bottom, the potential energy transforms into time-averaged kinetic energy and the turbulent kinetic energy increases. As seen from these results, the high-value turbulent kinetic energy region appears in the penstock and the test section. The maximum value of the turbulent kinetic energy is equal to $k = 4.02$ m^2/s with the standard k-ε model, $k = 3.57$ m^2/s with the realizable k-ε model, $k = 1.21$ m^2/s with the RNG k-ε model, $k = 20.61$ m^2/s with the standard k-ω model, $k = 1.70$ m^2/s with the transition-SST model and $k = 2.19$ m^2/s with the BSL k-ω model. In addition, it can be seen that the levels of turbulent kinetic energy increase in the downstream direction, with significant values appearing in the region closer to the free surface, and higher-energy regions moving toward the wall. Based on

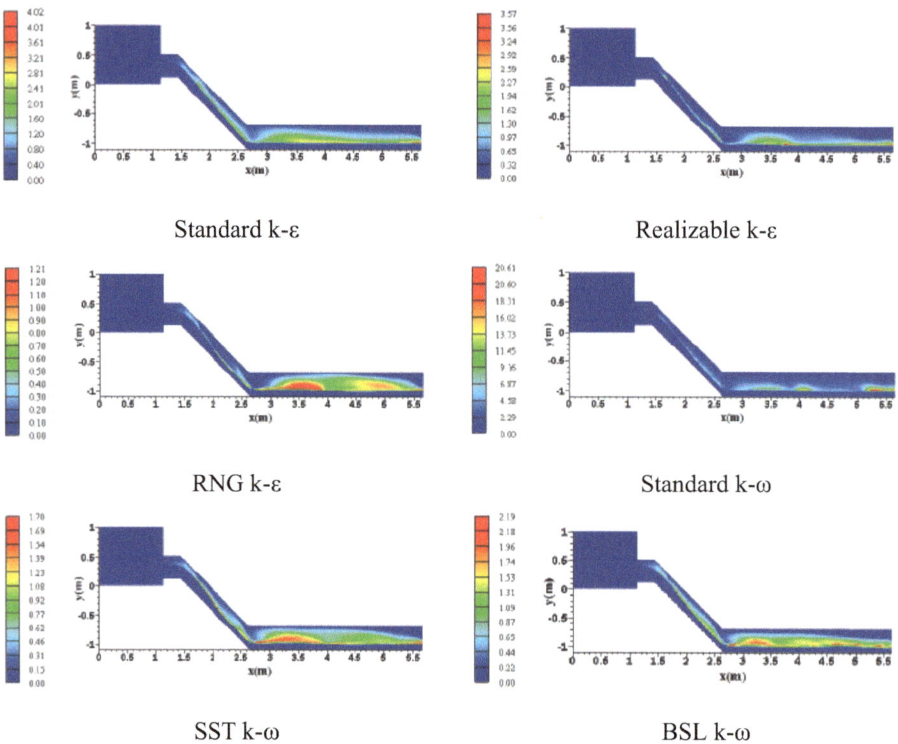

Figure 13.9: Turbulent kinetic energy distribution.

the comparison between the findings, we can confirm that the turbulence models have an important impact on the final solution of the turbulent kinetic energy field.

13.6.7 Dissipation rate of turbulent kinetic energy

The distribution of the dissipation rate of turbulent kinetic energy in the plane defined by $z = 0.45$ m and $x = 3.6$ m over the hydraulic test bench is predicted by the standard *k-ε* model, the realizable *k-ε* model, the RNG *k-ε* model, the standard *k-ω* model, the transition SST model and the BSL *k-ω* model, as depicted in Figure 13.10. From these results, it is observed that the dissipation rate of turbulent kinetic energy varies significantly from one turbulence model to another. The production rate of Reynolds stresses in the flow field involves the production of turbulent kinetic energy and, eventually, the dissipation rate of this turbulent kinetic energy. As can be seen, the highest values of the turbulent dissipation rate are situated on the penstock and the test section. Indeed, the maximum value of the dissipation rate is equal to $\varepsilon = 96.55$ m^2/s^3 with the standard *k-ε* model, $\varepsilon = 81.88$ m^2/s^3 with the realizable *k-ε* model, $\varepsilon = 46.38$ m^2/s^3

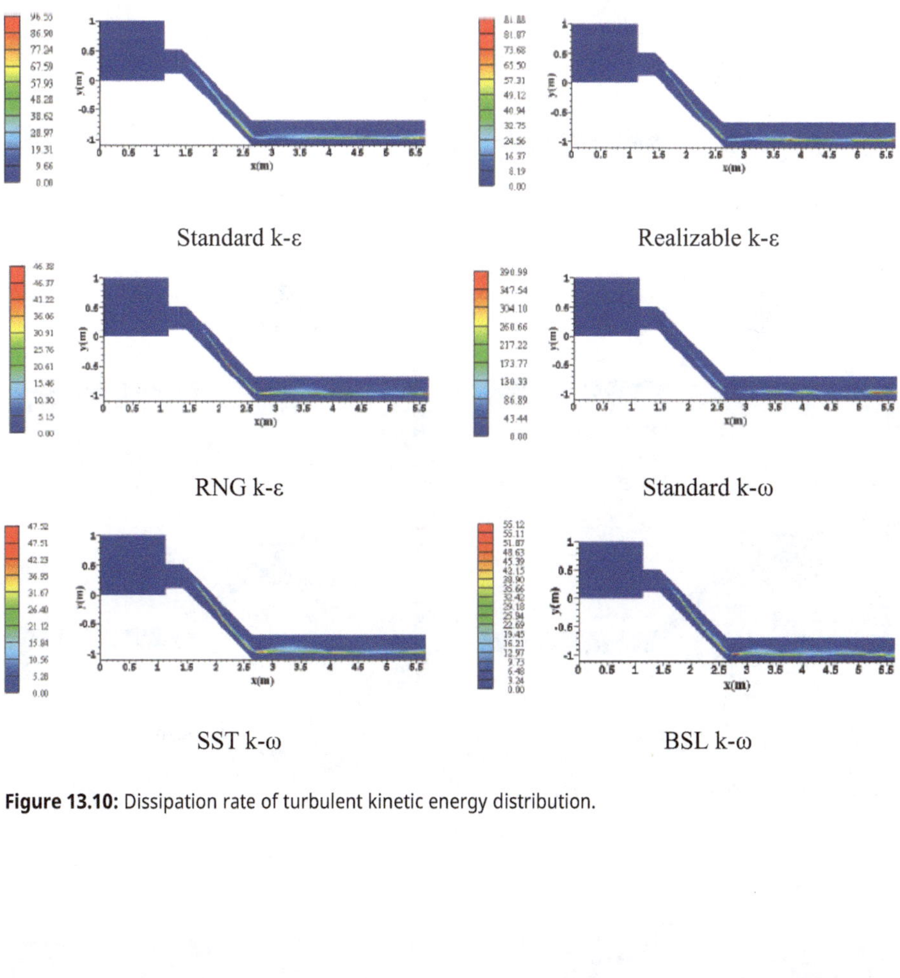

Figure 13.10: Dissipation rate of turbulent kinetic energy distribution.

with the RNG k-ε model, $\varepsilon = 390.99\ m^2/s^3$ with the standard k-ω model, $\varepsilon = 47.52\ m^2/s^3$ with the transition-SST model and $\varepsilon = 55.12\ m^2/s^3$ with the BSL k-ω model. The action of viscosity is strongest due to the large instantaneous velocity gradients that imply small-scale motions. Moreover, it can be observed that the turbulent kinetic energy is strongly dissipated near the free surface. The comparison between the results shows that the turbulence models have a significant effect on the final solution of the dissipation rate of the turbulent kinetic energy field.

13.6.8 Turbulent viscosity

Figure 13.11 depicts the turbulent viscosity distribution in the plane defined by $z = 0.45$ m and $x = 3.6$ m over the hydraulic test bench, predicted by the standard k-ε model, the realizable k-ε model, the RNG k-ε model, the standard k-ω model, the transition SST model and the BSL k-ω model. The turbulent viscosity does not depend on fluid features but depends only on the state of turbulence. Turbulence models are leaned on the Boussinesq assumption. They confirm that the momentum transfer resulting from turbu-

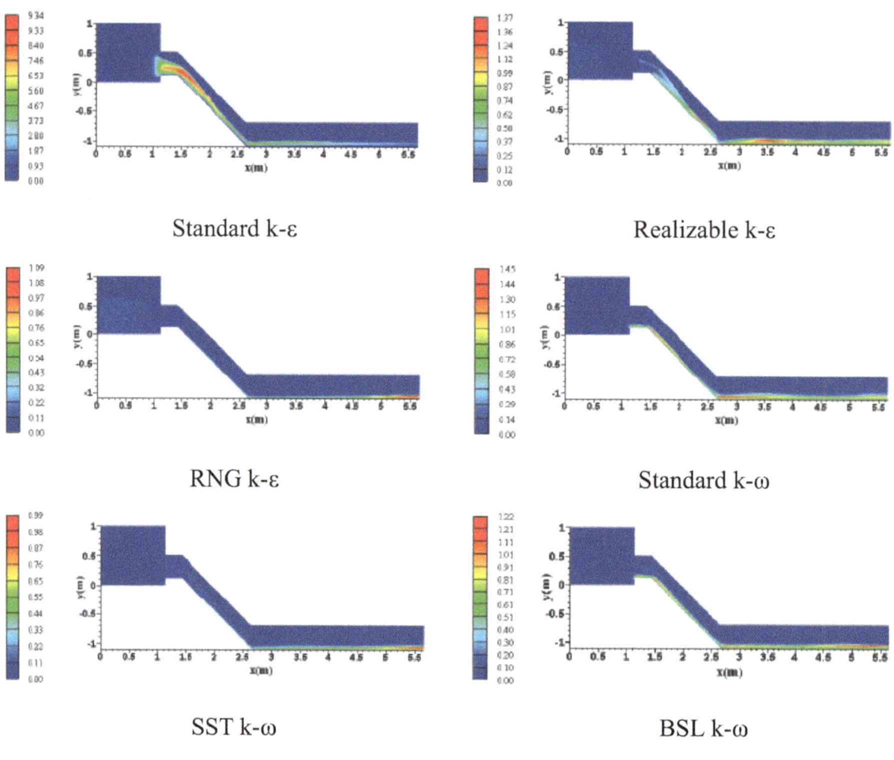

Figure 13.11: Turbulent viscosity distribution.

lent eddies can be modeled with an eddy viscosity. As can be perceived from these results, the highest values of the turbulent viscosity are achieved in the upper area of the penstock and in the test section, which matches the transient stage of a strongly anisotropic turbulent flow. Furthermore, the turbulent viscosity accomplishes its maximum value at the current head where the large eddies are produced. Indeed, it is clear that the of the turbulent viscosity is equal to $\mu_t = 9.34$ m^2/s with the standard *k*-*ε* model, to $\mu_t = 1.37$ m^2/s with the realizable *k*-*ε* model, to $\mu_t = 1.09$ m^2/s with the RNG *k*-*ε* model, to $\mu_t = 1.45$ m^2/s with the standard *k*-*ω* model, to $\mu_t = 0.99$ m^2/s with the transition-SST model and to $\mu_t = 1.22$ m^2/s with the BSL *k*-*ω* model. It is also obvious that the turbulent viscosity decreases in the direction of the flow. The turbulence models are of great importance. They have an important impact on the final solution of the turbulent viscosity field.

13.7 Conclusion

In this study, the performance of various turbulence models in predicting complex turbulent flow over a large-scale hydraulic test bench is assessed in detail. The six models selected for investigation are the standard *k*-*ε* model, the realizable *k*-*ε* model, the RNG *k*-*ε* model, the standard *k*-*ω* model, the transition SST model and the BSL *k*-*ω* model. The results of CFD computations were compared with the experimental measurement of [31] for the breaking of a dam. It is worth noting that the RNG *k*-*ε* turbulence model provided the best findings for the magnitude velocity profiles along the test section compared to the other models. We can conclude that the RNG *k*-*ε* model is the best in terms of accuracy, stability and efficiency of the computed solutions. It is recommended to use more complex physical models, including the LES for the purpose of resolving the details of the flow field.

References

[1] L. Davidson. An Introduction to Turbulence Models. 2022, 1–50.

[2] O. Darrigol and J. S. Turner Worlds of flow: A history of hydrodynamics from the Bernoulli to Prandtl. Physics Today, 59(8):52–54, 2006.

[3] J. Kim, P. Moin and R. Moser. Turbulence statistics in fully developed channel flow at low Reynolds number. Journal of Fluid Mechanics, 177:133–166, 1987.

[4] A. Rastegari and R. Akhavan. The common mechanism of turbulent skin-friction drag reduction with superhydrophobic longitudinal microgrooves and riblets. Journal of Fluid Mechanics, 838:68–104, 2018.

[5] S. S. Girimaji. Partially-averaged Navier-stokes model for turbulence: A Reynolds-averaged Navier-stokes to direct numerical simulation bridging method. Journal of Applied Mechanics, Transactions ASME, 73(3):413–421, 2006.

[6] P. R. Spalart. Detached-eddy simulation. Annual Review of Fluid Mechanics, 41:181–202, 2009.

[7] D. Laurence. Large Eddy Simulation of Industrial Flows? Closure Strategies for Turbulent and Transitional Flows. In: Host Publication. 392–406, 2002.

[8] K. Hanjalic Will RANS survive LES? A view of perspectives. Journal of Applied Mechanics, Transactions ASME, 127(5):831–839, 2005.
[9] J. P. Slotnick, A. Khodadoust, J. Alonso, D. Darmofal, W. Gropp, E. Lurie, et al. CFD vision 2030 study: A path to revolutionary computational aerosciences. 2014.
[10] M. Woelke Eddy viscosity turbulence models employed by computational fluid dynamic. Prace Instytutu Lotnictwa, 4(191):92–113, 2007.
[11] C. D. Argyropoulos and N. C. Markatos. Recent advances on the numerical modelling of turbulent flows. Applied Mathematical Modelling, 39(2):693–732, 2015.
[12] P. A. Durbin. Some recent developments in turbulence closure modeling. Annual Review of Fluid Mechanics. 2018;50:77–103.
[13] K. Duraisamy, G. Iaccarino and H. Xiao. Turbulence modeling in the age of data. Annual Review of Fluid Mechanics, 51:357–377, 2019.
[14] B. E. Launder. Second-moment closure: Present . . . and future? International Journal of Heat and Fluid Flow. 10(4):282–300, 1989.
[15] C. G. Speziale, S. Sarkar and T. B. Gatski. Modelling the pressure–strain correlation of turbulence: An invariant dynamical systems approach. Journal of Fluid Mechanics, 227:245–272, 1991.
[16] T. B. Gatski and C. L. Rumsey. Linear and nonlinear eddy viscosity models. Closure Strategies for Modelling Turbulent and Transitional Flows. 9–46, 2010.
[17] K. Hanjalić and S. Kenjereš. Some developments in turbulence modeling for wind and environmental engineering. Journal of Wind Engineering and Industrial Aerodynamics, 96(10–11):1537–1570, 2008.
[18] K. Suga. Recent developments in eddy viscosity modelling of turbulence. R&D Review of Toyota CRDL. 33(1):4, 1998.
[19] D. C. Wilcox. Turbulence Modeling for CFD. DCW Industries. Inc, La Canada, CA, November. 2006, 34.
[20] S. B. Pope. Turbulent Flows. Cambridge university press, 2000.
[21] P. A. Durbin and B. A. P. Reif. Statistical Theory and Modeling for Turbulent Flows: Second Edition. Statistical Theory and Modeling for Turbulent Flows. Second Edition. John Wiley & Sons; 2010.
[22] M. Leschziner. Statistical Turbulence Modelling for Fluid Dynamics – Demystified: An Introductory Text for Graduate Engineering Students [Internet]. World Scientific. 2015, 1–424. Available from: www.Ebook777.com
[23] J. O. Hinze. Turbulence (Chapter 5.7). McGraw-Hill, 1975.
[24] B. Launder The numerical computation of turbulent flows. Computer Methods in Applied Mechanics and Engineering, 3(2):269–289, 1974.
[25] T.-H. Shih, W. W. Liou, A. Shabbir, Z. Yang and J. Zhu. A new k-ϵ eddy viscosity model for high Reynolds number turbulent flows. Computational Fluid, 24(3):227–238, 1995.
[26] V. Yakhot and S. A. Orszag. Renormalization group analysis of turbulence. I. Basic Theory. Journal of Scientific Computing, 1(1):3–51, 1986.
[27] D. C. Wilcox. Formulation of the kw turbulence model revisited. AIAA Journal, 46(11):2823–2838, 2008.
[28] F. R. Menter. Two-equation eddy-viscosity turbulence models for engineering applications. AIAA Journal, 32(8):1598–1605, 1994.
[29] I. ANSYS. Fluent User's Guide 18.2. ANSYS, Inc, Canonsburg, PA, USA, 2017.
[30] C. W. Hirt and B. D. Nichols Volume of fluid (VOF) method for the dynamics of free boundaries. Journal of Computational Physics. 39(1):201–225, 1981.
[31] S. Koshizuka. Moving-particle semi-implicit method for fragmentation of incompressible fluid. Nuclear Science and Engineering, 123:421–434, 1995.

Index

https://doi.org/10.1515/9783110717655-014

De Gruyter Series Smart Computing Applications

ISSN 2700-6239
e-ISSN 2700-6247

Deep Learning for Cognitive Computing Systems
Technological Advancements and Applications
Edited by M.G. Sumithra, Rajesh Kumar Dhanaraj, Celestine Iwendi and Anto Merline Manoharan, 2022
ISBN 978-3-11-075050-8, e-ISBN (PDF) 978-3-11-075058-4, e-ISBN (EPUB) 978-3-11-075061-4

Cloud Analytics for Industry 4.0
Edited by Prasenjit Chatterjee, Dilbagh Panchal, Dragan Pamucar and Sharfaraz Ashemkhani Zolfani, 2022
ISBN 978-3-11-077149-7, e-ISBN (PDF) 978-3-11-077157-2, e-ISBN (EPUB) 978-3-11-077166-4

Advances in Industry 4.0
Concepts and Applications
Edited by M. Niranjanamurthy, Sheng-Lung Peng, E. Naresh, S. R. Jayasimha, Valentina Emilia Balas, 2022
ISBN 978-3-11-072536-0, e-ISBN (PDF) 978-3-11-072549-0, e-ISBN (EPUB) 978 3 11 072553-7

Soft Computing and Optimization Techniques for Sustainable Agriculture
Debesh Mishra, Suchismita Satapathy, Prasenjit Chatterjee, 2022
ISBN 978-3-11-074495-8, e-ISBN (PDF) 978-3-11-074536-8, e-ISBN (EPUB) 978-3-11-074537-5

Knowledge Engineering for Modern Information Systems
Methods, Models and Tools
Edited by Anand Sharma, Sandeep Kautish, Prateek Agrawal, Vishu Madaan, Charu Gupta, Saurav Nanda, 2021
ISBN 978-3-11-071316-9, e-ISBN (PDF) 978-3-11-071363-3, e-ISBN (EPUB) 978-3-11-071369-5

Knowledge Management and Web 3.0
Next Generation Business Models
Edited by Sandeep Kautish, Deepmala Singh, Zdzlslaw Polkowski, Alka Mayura, Mary Jeyanthi, 2021
ISBN 978-3-11-072264-2, e-ISBN (PDF) 978-3-11-072278-9, e-ISBN (EPUB) 978-3-11-072293-2

Cloud Security
Techniques and Applications
Edited by Sirisha Potluri, Katta Subba Rao, Sachi Nandan Mohanty, 2021
ISBN 978-3-11-073750-9, e-ISBN (PDF) 978-3-11-073257-3, e-ISBN (EPUB) 978-3-11-073270-2

www.degruyter.com